Springer Theses

Recognizing Outstanding Ph.D. Research

For further volumes:
http://www.springer.com/series/8790

Aims and Scope

The series "Springer Theses" brings together a selection of the very best Ph.D. theses from around the world and across the physical sciences. Nominated and endorsed by two recognized specialists, each published volume has been selected for its scientific excellence and the high impact of its contents for the pertinent field of research. For greater accessibility to non-specialists, the published versions include an extended introduction, as well as a foreword by the student's supervisor explaining the special relevance of the work for the field. As a whole, the series will provide a valuable resource both for newcomers to the research fields described, and for other scientists seeking detailed background information on special questions. Finally, it provides an accredited documentation of the valuable contributions made by today's younger generation of scientists.

Theses are accepted into the series by invited nomination only and must fulfill all of the following criteria

- They must be written in good English.
- The topic should fall within the confines of Chemistry, Physics, Earth Sciences, Engineering and related interdisciplinary fields such as Materials, Nanoscience, Chemical Engineering, Complex Systems and Biophysics.
- The work reported in the thesis must represent a significant scientific advance.
- If the thesis includes previously published material, permission to reproduce this must be gained from the respective copyright holder.
- They must have been examined and passed during the 12 months prior to nomination.
- Each thesis should include a foreword by the supervisor outlining the significance of its content.
- The theses should have a clearly defined structure including an introduction accessible to scientists not expert in that particular field.

Volker Zepf

Rare Earth Elements

A New Approach to the Nexus of Supply, Demand and Use: Exemplified along the Use of Neodymium in Permanent Magnets

Doctoral Thesis accepted by
the University of Augsburg, Germany

Springer

Author
Dr. Volker Zepf
Coordinator Graduate Programme
Augsburg University
Augsburg
Germany

Supervisors
Prof. Dr. Armin Reller
Chair of Resource Strategy
Department of Physics
Augsburg University
Augsburg
Germany

Prof. Dr. Gerd Peyke
Chair of Human Geography and Geoinformatics
Augsburg University
Augsburg
Germany

ISSN 2190-5053 ISSN 2190-5061 (electronic)
ISBN 978-3-642-35457-1 ISBN 978-3-642-35458-8 (eBook)
DOI 10.1007/978-3-642-35458-8
Springer Heidelberg New York Dordrecht London

Library of Congress Control Number: 2012955237

Printed on acid-free paper

Springer is part of Springer Science+Business Media (www.springer.com)

Supervisor's Foreword

Some of the globally implemented technologies which were developed and invented in the last two and three decades often rely on the properties of appropriately selected functional materials. In most cases these materials contain or are metals themselves, alloys or metallic compounds. This trend of increasing diversity of metals is becoming obvious above all in IT, mobility, and energy technologies. Concomitantly to this increasing qualitative material diversity the specific material quantities in products like computers, digital cameras, mobile phones, wind turbines or electric motors differ extremely. Very often, the quantities of the functional materials are very small but indispensable for the main functionality.

The geologic availability of the mineral resources which are necessary for providing metallic components is partly critical, especially in the case of trace or spice metals. Partially extreme dynamics, under which new technologies and novel product generations are implemented, make it understandable that supply bottlenecks arise. Resource shortage, i.e. scarcity becomes imminent.

Such undesired and threatening situations particularly occur when multifunctional metals are functionalized in different technical branches or products. The 17 elements of the so-called rare earth metals unmistakably belong to this type of critical materials. They show a huge bandwidth of application areas and serve as essential functional materials, alloys, and ceramics. The economic trade restrictions imposed by China exemplary show the potential of and for supply shortages. Thus the rare earth metals can act as a paradigm for the present global resource situation and the impact on the industrial development worldwide. This prevailing situation is the basis for the doctoral thesis of Mr. Volker Zepf.

The first introductory part of the thesis presents a detailed analysis of the history, geology, chemical, and physical characteristics of the rare earth metals. As well socio-economic, cultural and political facets are addressed which show the partial indispensable technical functionality of these metals. The thorough and embracing research on technical-historical functionalities of the rare earth metals and their practical relevance lead Mr. Zepf to a critical summary: valid and reliable primary data about supply and use of rare earth metals is extremely difficult to attain despite the many publications in recent years. The data used therein can mostly be traced back to one literature source—and the data used often is unreliable.

The main investigation now aims on providing reliable data. The focus is narrowed on the use of neodymium functionalized in permanent magnets in computer hard drives, mobile phones, wind turbines, and electric motors in e-mobility. With respect to several headlines the hypothesis is raised that these four application areas account for about 80 % of the global neodymium demand. The analysis then provides reproducible primary data for selected application areas. Here astonishing and surprising results are provided which refute the original hypothesis by large; only about 20 % could be attributed. This result now is interesting in several aspects: it shows that a wrong perception prevails on the use of neodymium. It also raises the demanding question on the other uses of neodymium nowadays. This knowledge is indispensable when projecting potential supply risks and the potential for re-phasing rare earth elements, i.e. re-cycling.

Finally a resource-geographical consideration combines the results in an impressive and comprehensive global materials chart. This chart can build the basis for further work and especially in the evaluation of potential secondary or urban rare earth mining. In how far the newest economic but also economic developments could shape our technically oriented civilization is subject of the closing multifaceted discussion.

Mr. Zepf has demonstrated an extremely high motivation, initiative, and competence with his thesis about a highly topical area. For this Mr. Zepf provides a multidisciplinary view and methodological approach. He develops a multi-structured and complex linked but still clearly structured, consistent, and relevant study. This thesis is of essential scientific relevance as it reveals unknown facts around the supply, demand and use of rare earth metals which is of great importance for further research and economic considerations especially for recycling options.

Augsburg, Germany, January 2013 Prof. Dr. Armin Reller

Acknowledgments

Such a work, even though it is only around 150 pages long, is a lot of work which is not manageable without the help of a lot of people, involved directly or indirectly.

First of all I thank Silvia for her extremely good, enduring, and motivating support during all the time of this work. Without you and your strength I would not have been able to do this work!

Next, my gratitude is to Prof. Dr. Armin Reller, Chair of Resource Strategy, Augsburg University, who gave me this interesting and challenging topic. For his visionary ideas and his quest for revealing links, relations, interrelations in an inter- and transdisciplinary approach. He gave me much freedom for my research and provided the decisive support when necessary; especially for getting the possibility to have the samples analyzed by the Fraunhofer Institute ISC.

I like to thank Prof. Dr. Gerd Peyke, Chair of Geography and Geoinformatics, Augsburg University, who always believed in me and for his interest in all the various aspects as well in the overall picture and for the discussions over the years.

Benjamin Achzet helped me with his expertise in materials science and how to start analyzing materials.

A special thank you goes to the Fraunhofer Institute ISC in Würzburg, and especially to Dr. Römer who did and observed the analytic measurements of the magnets.

And last but not least thank you all: the ones I forgot and the diligent people at the University and from Infau, Augsburg, who helped and provided me with hard disks, mobile phones, and other electric tools.

Thank you!

Contents

Abbreviations, Units Used, and Citation Note

These abbreviations, mainly the ones for the 17 rare earth elements are used in this Book.

AAGR	Average annual growth rate
BGR	Bundesanstalt für Geowissenschaften und Rohstoffe
BGS	British Geological Survey
CAGR	Continuous annual growth rate
Ce	Cerium
CREIC	Chinese Rare Earth Information Center, sometimes just referred to as CREI
CSRE	The Chinese Society of Rare Earths
Dy	Dysprosium
EDX	Energy dispersive X-ray analysis
Er	Erbium
Eu	Europium
Gd	Gadolinium
GW	Giga Watt
GWEC	Global Wind Energy Council
HDD	Hard disk drive
Ho	Holmium
HTS	High temperature superconducting (envisaged future wind energy technology)
IDC	International Data Corporation, market intelligence, and advisory services
La	Lanthanum
Ln	Lanthanoids, also called Lanthanides
Lu	Lutetium
MCS	Mineral Commodity Summaries, issued by the USGS
MGOe	Megagaussoerstedt
MIIT	(Chinese) Ministry of Industry and Information Technology
MLR	(Chinese) Ministry of Land and Resources

MOC	(Chinese) Ministry of Commerce
MOST	Ministry of Science and Technology of the People's Republic of China
MW	Mega Watt
MYB	Minerals Yearbook, issued by the USGS
Nd	Neodymium
NdFeB	Neodymium, iron, boron (magnet)
NDRC	National Development and Reform Commission (China)
Pm	Promethium
Pr	Praseodymium
REE	Rare earth elements
REM	Raster electron microscope
REO	Rare earth oxide
REPM	Rare earth permanentmagnet
REPMG	Rare earth-based permanent magnet generator (wind turbine type)
RFA	Röntgenfluoreszenzanalyse (=XRF—X-ray fluorescence analysis)
Sc	Scandium
Sm	Samarium
SSD	Solid state disk
t, ton	metric ton; 1,000 kg
Tb	Terbium
Tm	Thulium
USGS	United States Geological Survey
VCA	Voice coil assembly; also used synonymously with VCM
VCM	Voice coil motor; also used synonymously with VCA
WD	Western Digital (company)
WTG	Wind turbine generator, also called wind turbine
Y	Yttrium
Yb	Ytterbium

Units

Units Metric system

Decimal dots and thousand dividing commas have been used

Chapter 1
Introduction

Rare earth elements (REE) have gained enormous economic, public and increasingly scientific interest in the few recent years. Numerous headlines have appeared during the last about 4 years dealing with REE in the news and newspapers around the world, in press announcements, as short informational letters as well as numerous reports and studies from consultants, think tanks and governmental research groups. The vastness of the reports could lead to the conclusion that all has been said about REE, that the problems have been identified and the actions to be taken are known. Even though REE are not traded on the stock exchanges there is a market which is basically controlled by Chinese traders. The prices of REE were for some time modest until within the last 2 years the prices drove rollercoaster. Extreme price increases as well as reductions were paired with lots of uncertainties about the reasons for these price jumps. It was and probably still is unsure what the real reasons were: real physical shortages or maybe even oversupply, actual or pretended stockpiles, politics, trade or the hope for fast and big money? The fear is further aggravated by studies around the world about rare metals and rare elements which identified REE as some of the most critical materials nowadays and in the near future. As China has a de facto monopoly of the REE production and is successively restricting export, fingers are pointed on China for restricting the global supply of REE and for not doing business right.

The topic of the REE got momentum in recent years with the call for ever more climate friendly systems like energy saving lamps, electric vehicles or wind turbines. Most of which can and partially must be produced with REE based materials like phosphors or permanent magnets (REPM) to achieve a reduction in size, weight and thus attain a high efficiency. This fact alone should not have gotten that much attention to the REE as they actually got. Responsible was the de facto monopoly of China as basically the only REE producer globally since several years. There not only the fact of the monopoly itself, but rather the increasing influence and control taken by the government attracted attention and fears. The absolute dependence on China for the supply of REE not only alarmed industries and governments around the globe but as well the US military which sees an

V. Zepf, *Rare Earth Elements*, Springer Theses,
DOI: 10.1007/978-3-642-35458-8_1,

unacceptable dependence on China for the homeland defense. Maybe it is not only the military which sees unwanted dependencies but also the defense industries. Several cases were brought forward to the US government which contributed to the urgency of the REE issue.

Now this Chinese monopoly situation will most probably change as new mines in Australia, the US and in some other countries are supposed to come in operation within the next few years. However on the stock exchanges the shares of these mining companies showed erratic and extremely volatile price behaviors; despite the fact, that earnings are visible on the horizon. This observation might be reduced to financial gambling actions and is probably not surprising. But this whole situation gets a strange touch when the static range of REE is taken into consideration. According to the data disseminated by the U. S. Geological Survey (USGS) the static range is a little more than 850 years.

Of course this range figure of 850+ years stands in strong contradiction to the identified and issued criticality of REE.

So more questions should arise and have to be asked like: Why is there a shortage when there is an appeasing buffer of 850 years? Or asked another way around: are the REE really that rare? If so—and even if not—, what impact does it have on mining and supply?

Getting away from the shortage issue towards a more general view of the REE: how can we compare elements used for the glass industry with such that are used for magnets fabrication, i.e. cerium with neodymium? And if we put and talk about them together as one, how does that influence the problematic, the reporting, understanding and the solutions? And who is it that provides all the information?

All the mentioned reports, studies, papers and insights give only partial answers. They show consistencies and a lot of information, but also some discrepancies, inaccuracies, misunderstandings and misconceptions. The reason for this is seen as an inherent problem of the issue about the REE itself: the sheer quantity of the options, relationships, similarities as well as some absolutely unique characteristics of the single elements, which often are summarized under the term 'REE', and thus are treated like one element.

The REE are a group of 17 elements which share some fascinating similarities but at the same time present quite some extremely different characteristics, parameters and behaviors which do not at all allow a one-to-one substitution between the individual elements. Instead these many different characteristics allow a huge bandwidth of applications ranging from lighter flints (cerium), UV glass protection (cerium), magnets for electric vehicles (neodymium, praseodymium, dysprosium and terbium) to NiMH batteries using lanthanum, yttrium for YAG lasers or erbium for signal amplification in glass fiber cables. One more argument shall be added to the already mentioned complexity which is the 'rareness', so to speak the abundances of the REE in the earths' crust. It is said that REE are not as rare as the name might imply. In fact the abundance ranges from 68 ppm for cerium (as abundant as copper) to 0.48 ppm for thulium, which is still a factor 100 more abundant than gold (acc. to data of [2]). This suggests to consider the REE in their individual appearance rather than talking about them as a homogeneous group or as 'one' element. In order to make the task workable and make the complexity manageable a reduction from the REE as one group to the consideration of functional groups may be possible, e.g. consider the permanent magnet elements together, i.e. praseodymium, neodymium, samarium, dysprosium and terbium. Even better would be the look at the individual elements alone. In this work, the reduction or emphasis will be put on Neodymium, as there was the most data available.

The real problem finally is indeed an information deficit and an unbalanced perception; mainly based on incomplete and partially unreliable data. This is despite the fact that there are numerous reports around. It will be shown later in this work that the supply data originates in principle from one source which is under scientific aspects neither reliable nor can the data be proofed for validity. So the hypothesis is that unreliable supply and demand figures build the basis for the discussion around the REE. It is also questionable whether in this case supply and demand are balanced or if demand actually represents real use of REE. The aim therefore is to contribute to this unsatisfactory situation by bringing some transparency into the demand or better production data. To accomplish this, a bottom up approach was selected starting with end user products rather than using and relying solely on published production data. End user products, i.e. HDDs, mobile phones and earphones have been dismantled and REE based magnets extracted, measured and analyzed. It is known and accepted that from a strong statistical point of view still deficiencies exist, however scales and relationships could be determined with unexpected results: the investigated products HDDs, mobile phones, wind turbines and electro motors for e-mobility do not play the major role of Nd use. Of course this result opens the field for further research. Finally the sequence of events in the REE case can teach some lessons which should be taken by heart so that respective measures will be drawn.

This chapter concludes the introduction with the motivation that stands behind this work. It will also unfold the informational dilemma that the REE issue is in and it gives a brief description of the methodology used. In Chap. 2 the REE are explained in some detail as experience showed that despite a lot of reports and publications still some misunderstandings are present. These can and do lead to misperceptions which are problematic for a factual reflection of the REE. First, some definitions and characteristics are described, followed by a brief history of the discovery of the REE. Then an introduction into the geology and geochemistry is given which eventually can explain potential deposits and thus present and future mining projects. The REE processing is explained as it is responsible for a variety of environmental problems and the processing and separation is one of the more challenging parts in the product cycle of the REE. The chapter ends with the applications of the REE, considered from different aspects. Chapter 3 contributes some insights into the Chinese situation which is deemed absolutely necessary to know in order to understand the Chinese behavior and actions. Chapter 4 now details one major point of this research: a literature study of selected REE reports which tries to reveal the actual data sources of the REE production data. A further topic in this literature research highlights the data discrepancies which are present but often are not obvious. Here also a first approach is made towards the most probable or reliable production data, both for the REE in total, along their functional areas and finally for Neodymium alone. These figures will later be used for the comparison of relationships and scales of Nd uses. In Chap. 5 a very short introduction into the permanent magnet area is given before in the next chapter the basic research data will be explained: the disassembly and measurement of HDDs and mobile phones plus some earphones and the literature research and derivation of production data of REE in wind turbines and electro mobility. The synopsis of the REPM use will eventually lead to a short summary of the geography of the REE where a map will be assembled to show the REE along space and time on a world map. This map shall show the life cycle of the REE and shall act as a starting point for evaluating the potentials for re-phasing, e.g. recycling. The last chapter concludes the observations made with a summary and some hints for further research. In the annexes detailed tables of the literature research are given as well as the lists of analyzed HDDs, mobile phones and a list of uses along the individual REEs.

1.1 Motivation

The basic topic of this work, the Rare Earth Elements, was suggested by Prof. Dr. Armin Reller, Chair of Resource Strategy, Augsburg University in 2009. At that time the REE gained increasing media attention and their importance got into focus in several US and EU studies. The unique properties of the REE seemed to be indispensable and ideally suited for climate friendly uses, i.e. the use in regenerative energy systems and emission-free mobility. The studies thus earmarked the REE as some of the most critical elements nowadays. The reason was mainly related to the monopolistic situation of China in the production of the REE, from mining to the component production like magnets or lamps; even though that fact was already known—and accepted—since years. This situation alone could be alarming enough, however the situation aggravated as China increased their efforts on controlling the REE production and export chain. Production and export restrictions were set up and taxes were raised. The 'west' now claimed unfair behavior combined with the fear of not getting the important raw materials for the praised green energy technologies to challenge climate change. Headlines and reports jumped on this prominent topic; commodity prices showed dramatic increases; ideas about substitutions got known, engagement of the producing industry in mining ventures emerged and finally claims against China were set by the West.

In this context initially everything seemed to be clear and obvious. However, after several months of research it showed that this assumption was not at all that clear. Several data sets seemed incorrect or didn't match with other sets. The static range derived from USGS data resulted in more than 850 years of supply, but there is a shortage today. How can that be?

A closer look at several reports and studies showed that the authors usually were aware that it is not easy to get reliable data. On the other hand there is a lot of very in depth insight and deep scientific knowledge in several REE topics like magnetism, wind turbine design, chemical processing of REE etc.

So the motivation for this work arose from these thoughts and contradictory information. The idea was to contribute some more reliable facts about REE production, demand and use data with this work. Eventually a basic research work evolved using a bottom up approach where products were disassembled, measured and analyzed in the hope to provide valuable and verifiable data. With these data at least some of the uncertainties could be clarified, some questions could be answered, but also new research questions have been identified.

1.2 An Information Dilemma

Since about 2009 numerous studies, reports and papers about REE have been published, mainly from consultants (e.g. [3, 4, 6, 7]) and research institutes (e.g. [8, 9]). These were accompanied by innumerable news articles around the world. All of them are similar in contents, figures and data with only moderate differences in estimations about possible future mining activities. Usually a gloomy picture is drawn showing the absolute dependence on China, on its will for production and export, respectively the increasing regulation of export. The conclusion drawn and implied usually is that mainly the buildup of green economies is endangered, i.e. the production of electric vehicles, wind turbines but also the computer hard drives and mobile phones are at stake. This impression is widespread and has been heard in numerous personal discussions in the past months. The argument for the criticality is seemingly fed by enormously increasing commodity prices for REE, especially in the first half of 2011. It also has to be said, that in the second half of 2011, the prices declined considerably again. All in all there are very volatile prices which of course are difficult to deal with. At the same time another tenor can be heard that REE are not as rare as the name implies and that reserves are huge enough to last for hundreds of years into the future. The USGS [10] announces reserves data with 110 Mio. t which result in a static range of about 850 years (110 Mio. t reserves divided by an annual production volume of 130,000 t in 2010). So why is there this huge contradiction? Is there a problem or not or where actually is the problem?

The literature and data source analysis will show that most or all data originate from one source but yet diverging data are published. Several authors are aware of the data reliabilities and state that their data is estimated and based on discussions with stakeholders. These discussions of course are no scientific expert interviews with transcriptions so that the contents cannot be verified. Obviously the application of intense scientific criteria is sometimes neglected in favor of more common sense data; probably due to time constraints and proprietary information issues. But if this is the case, there is no reliable and valid data available. Indeed the many data sets, reports and continuing headlines imply a sound knowledge of the issue, but actually there is a lot of uncertainty and vagueness. But that is not apparent. In fact it has to be attested that there exists an informational dilemma. Only very few papers deal with the common misconceptions like e.g. Hurst [5].

Probably all information or 'facts' contain some truth but the conclusions drawn are not based on reliable data and thus there is a mismatch and misunderstanding of the real scales, relationships and the magnitude of the individual topics like 'REE in wind energy' etc. The result is that a wrong perception of the real problem can and does happen. Stosch [11] already mentions in 2002 that the experience of recent years showed an over-interpretation of REE data. Out of this informational and perceptional dilemma a distorted knowledge arises which eventually leads to conclusions and actions which are actually based on wrong assumptions, whereas the real problem is not obvious.

This work will show which relational share the headline applications HDDs, mobile phones, wind turbines and e-mobility have in the global annual production. It is often assumed that the REE are indispensable in these technologies—which is only partially true. And it will be shown that next to these four technological areas considerable numbers of other applications require REE. These are not in the focus but they may play a vital and decisive role in determining the real magnitude of REE demand in future.

Maybe the 'problem' with the REE, namely a potential short term shortage in supply, is resolved in 5–10 years' time. Still a lesson can and has to be drawn from the 'story of the REE'. This could serve as an example on how to act or better on how not to act in future concerning minerals and metals handling, both for individuals, companies and governments, i.e. giving up mining activities and with it know how in the succeeding production chain. This behavior is getting a special touch when it is done with the knowledge that this could result in a global monopoly situation (the Chinese Premiers announced the importance of the REE for China and the world publicly and repeatedly). Eventually time proved exactly that importance. The question remains if China has done this on purpose, to lure the West in the trap, or whether the buck can be passed partially to the West in turn. Here no statements and evaluations of guilt or responsibilities will be made, instead these arguments and questions shall illustrate the complex situation and still the inaccurate picture of the REE issue. In the following chapters these individual topics will be addressed in more detail and it is the intention to solve at least some facets of the informational dilemma.

Eventually the leading questions for this work were:

What are the real quantities of Nd required to produce the HDDs, mobile phones, wind turbines and electro motors in one year and within the last about 10 years?

What are the correct annual production numbers of REE and Nd in particular?

Out of the answers to these questions further ideas can be deducted:

Is there actually an imminent or mid- to long-term shortage problem concerning the REE?

Do the four application areas under research play the major role in todays' Nd use? What are the consequences if they are, and what, if they are not?

Which methodology is suitable to provide the required data reliability and validity?

1.3 Methodology

1.3.1 Introductory Thoughts: Managing the Complexity

In an article about research trends in rare earths, Adachi et al. [1] had a look at the papers registered as Chemical Abstracts in the American Chemical Society since 1990. At that year 18,000 hits were counted for rare earth topics overall. In 2007, about 30,000 papers overall were noted. These papers were sorted and analyzed along 20 different application fields like magnet, separation, spectroscopy, battery and more. In the magnet application field for example about 3,000 published papers were counted for the years 2006–2008 each. Next to the sheer number of publications especially astonishing for the authors was that China progressed in RE research, simply due to their immense research funding and creation of specialized research centers [1]. Adachi's work gives a short impression about the vast field of the topic. It is a challenge to handle the information density in one application field alone, and it is nearly impossible, to manage the entire bandwidth. Another conclusion can be drawn as well, that there is a lot of expertise in the specialized fields themselves. What is missing however is a relational quantitative overview which reveals links, relationships and dependencies between the different (scientific) disciplines and the respective need for raw materials. The intent to contribute such a missing link is not trivial at all of course. It buries the danger of both digging too deep into one facet but neglecting the other but also not to dig deep enough to get to the essence. It should also be obvious that one person alone cannot know all details of all disciplines. Having this in mind I want to emphasize that it is not at all my intent, to claim knowledge in all disciplines. Very often I was reduced to literature study and did not have the chance, time and funding to get in touch with specialists, experts and stakeholders. So it is likely, that there are inaccuracies present in this work. These certainly are my faults, but they could also act as initiation of corrective notes to me and as a starting point for further clarifications and possibly further research. Some areas have been identified for further research and are annotated throughout the work. From a scientific methodological point of view it is a challenge as well, as I—as a geographer—do not at all have the specialized insight in all relevant disciplinary methodologies associated with the topic. But I see it is a task of a geographer and resource strategist to take the challenge and show relationships by using a combination of methodologies which are borrowed from mathematical-physical sciences like REM and RFA analyses, as well as there were methodologies used which are more common in arts and socio-economic sciences. A simple literature and source review turned out to be one key point because the data source analysis disclosed a problem that seemingly a lot of information and data about the topic exists. A closer look revealed that actually most, if not all, of the disseminated data stems actually from one data source only.

1.3.2 Methods and Data Used

It probably is not surprising that accurate data is more than difficult to get. This has been described by other authors as well. Such the underlying methodology is that of a basic research data sampling rather than the posture of a thesis with the intent of falsifying it. The methodology used aims at the support of the determination or better what share the Nd use actually has on the areas HDD, mobile phones (iPhones), wind energy and e-mobility. If you will it could yet be postured as the thesis 'the four areas HDD, mobile phones, wind energy and e-mobility quantitatively account for about 80 % of the annual production of Neodymium' with the intent to falsify this thesis. Surely this methodological topic could be elaborated in more detail but it will be set aside for this work. Even though this basic research approach might look like a step back—as some reports already think about substitutions of REE—it is seen as absolutely indispensable to have solid and reliable data before further steps are being taken.

For the overall picture about the topic of the REE, their geology, geo-chemistry, characteristics, history, and some basic definitions, an encyclopedic search through standard educational literature across several scientific disciplines has been performed to assemble sound and reliable information. In a next step, the status quo of the REE concerning production data was initiated with the obligatory literature research which was both done by scientific printed literature but also to a big extent by internet data sources. The topic of REE is extremely volatile and a lot of information is only provided online and as PDF files. Among these information both reports from national research agencies were involved, like the USGS, BGS, BGR, and from research institutes like the Öko-Institut, but also papers from consultants like Roskill or Byron. Eventually several reports and press releases from news agencies were used which inform especially about environmental problems expected in China.

All of this data however is considered as secondary data as the primary data could not be verified along official data like export lists or annual reports of producing companies.

As a central part of this research work a primary data inquiry has been done by an in depth research or bottom up approach, i.e. the disassembly of consumer products and the determination of actually used REE within these products. Therefore HDDs and mobile phones have been collected, dismantled and analyzed with REM-EDX[1] and RFA analyses. This is seen as a starting point for verifiable and reliable data.

[1] REM—Rasterelektronenmikroskop (scanning electron microscope); EDX—Energy Dispersive X-Ray Analysis; RFA—Röntgenfluoreszenz-Analyse (X-Ray Fluorescence Analysis).

For the further determination of global quantities next to these primary data also secondary data from market research agencies had to be used which provided global production data of the selected products. This fallback was necessary as not all companies disseminate production data in measures of units, but rather in financial results only. So there was no other way in the time and constraints given, as to rely on those analyst data. This of course buries some potential inaccuracies again. The quantities determined will show that some inaccuracies however will not change the message of this work.

For the area of wind energy most data was taken from annual reports of the companies as this data is considered as best available. This was uncomplicated for Western companies, but difficult for Chinese companies. They often present some basic data on their homepages or just by means of some World Wide Web informational platforms. More potential reasons for slight discrepancies lie in the consideration of fiscal versus calendar year data and the not standardized usages of attributes like sold, produced, shipped or installed wind power.

It was planned to do expert interviews and use questionnaires to get information about REE used in selected products. However, no returns of RFI were noted or no addressee could be found. During private discussion with stakeholders I got the impression that companies refrain from giving data away—not to speak from accurate data at all; probably because they fear competition disadvantages. As well the access to experts in China was not possible in the constraints given so that I refrained from the use of expert interviews and RFIs.

One last remark shall be added that several reports could not be used because they were too expensive. Several consultants' reports cost in the order of several thousand US $ so that the price restricts the access. It would have been of interest, which data sources these consultants use, and if they come up with the same or more reliable data as the freely available reports.

Even after this short introduction about methodology it should have become obvious that no single dedicated methodology is solely applicable, rather a mixture of methods has to be applied. The results of this work however are promising and should encourage respective successive projects and further research.

The topic of the REE is extremely dynamic and reports, papers and new numbers are published nearly weekly. In order to come to an end with the analysis the data used was about as of end of 2011.

1.4 Conclusion

The available data about REE does not always withstand scientific criteria. Even though this work again has to use some of such data derived from literature and data analyses, a starting point for another perspective has been set: the disassembly of seemingly decisive products to investigate for actual qualities and quantities of Nd use in selected products. Some prior scientific suggestions, ideas and results could be verified; others could be updated and improved. This work is of course not exhaustive and calls for a succession. For that more time, manpower and funding would be necessary.

As an accompanying task a scientific methodological discourse can be addressed: the problem of an inter- and trans-disciplinary approach that does and cannot satisfy all classic methodologies of the involved disciplines. Any possible shortfalls need discussion with the aim of the eventual scientific acceptance of methodologies and suitable methods to approach such a task.

References

1. Adachi G-Y, Imanaka N, Tamura S (2010) Research trends in rare earths: a preliminary analysis. J Rare Earths 28(6):843–846. doi:10.1016/S1002-0721(09)60207-6
2. Binder HH (1999) Lexikon der chemischen Elemente: Das Periodensystem in Fakten, Zahlen und Daten. S. Hirzel, Stuttgart
3. Byron (2010) Electric metals. Green book. Apr 2010. Retrieved from http://www.google.de/url?sa=t&rct=j&q=byron+electric+metals+green+book&source=web&cd=2&ved=0CDoQFjAB&url=http%3A%2F%2Fwww.byroncapitalmarkets.com%2Fwp-content%2Fuploads%2F2011%2F02%2FElectric-Metals-Green-Book-April-2010.pdf&ei=zswjT5_DOIer-QaCi73CCA&usg=AFQjCNEcMvN5AUPbxtFdd6VcNxwGexF4sA&cad=rja. Accessed 11 Nov 2012
4. Cormark (2011) Rare earth metals: critical to energy efficient products, mobile electronics, and electric vehicles. Investor information available and retrieved from http://de.slideshare.net/RareEarthsRareMetals/cormark-securities-rare-earth-metals-initiating-coverage. Accessed 11 Nov 2012
5. Hurst C (2011) Common misconceptions of rare earth elements. Retrieved from http://www.ensec.org/index.php?option=com_content&view=article&id=290:common-misconceptions-of-rare-earth-elements&catid=114:content0211&Itemid=374. Accessed 11 Nov 2012
6. Lifton J (2009) The rare earth crisis of 2009. Detroit. Retrieved from http://www.techmetalsresearch.com/2009/11/the-rare-earth-crisis-of-2009-part-2-the-green-wind-blows-from-china/. Accessed Nov 11 2012
7. Roskill (2007) The economics of rare earths and yttrium. 13th edn.
8. SATW (2010) Seltene Metalle: SATW Schrift 41, Nov 2010. Retrieved from http://www.satw.ch/publikationen/satwinfo/SelteneMetalle.pdf. Accessed 11 Nov 2012
9. Schüler D, Buchert M, Liu R, Dittrich S, Merz C (2011) Study on rare earths and their recycling: final report for the greens/EFA group in the European parliament. Darmstadt. Retrieved from http://www.oeko.de/oekodoc/1112/2011-003-en.pdf. Accessed 11 Nov 2012
10. Stosch HG (2002) Geochemie der Seltenen Erden: Skript zur Vorlesung am Mineralogisch-Petrographischen Institut der Universität zu Köln, 1988–1993, Skript mit Ergänzungen von 1998 und Sommer 2000. Retrieved from http://institut-seltene-erden.org/wp-content/uploads/2012/03/uni-koeln-geochemie-der-seltenen-erden.pdf. Accessed 11 Nov 2012. Newer versions are available but not directly retrievable on the internet
11. USGS (2011) Mineral commodity summaries. Rare earths. Jan 2011. Retrieved from http://minerals.usgs.gov/minerals/pubs/commodity/rare_earths/mcs-2011-raree.pdf Accessed 11 Nov 2012

Chapter 2
Rare Earth Elements: What and Where They Are

2.1 Definitions

REE are often misunderstood, beginning from the grouping of the relevant elements to the etymological misconception, that REE are rare or the fact that REE are 'more abundant than gold'. Thus REE can get the impetus of being very abundant, which is only part of the truth. As well the allegation "The term rare earth is actually a misnomer" [1, p. 3] shows a cursory view of these metals. This view is for most purposes probably good enough, like the need for a 5-min-overview of the REE. The problem here lies in a possible wrong perception which finally gets to a seemingly hard fact because 'everybody says so'. In order to get a scientific plausible picture, several definitions are given at the beginning.

2.1.1 The Group of the REE

The Rare Earth Elements (REE), also simply called Rare Earths (RE) or Rare Earth Metals (REM), are a group of 17 elements—according to the International Union of Pure and Applied Chemistry (IUPAC) [2, p. 51]. The 17 REE therefore consist of the elements scandium (Sc), yttrium (Y) and the 15 so called lanthanoids (Ln) which are the elements lanthanum (La), cerium (Ce), praseodymium (Pr), neodymium (Nd), promethium (Pm), samarium (Sm), europium (Eu), gadolinium (Gd), terbium (Tb), dysprosium (Dy), holmium (Ho), erbium (Er), thulium (Th), ytterbium (Yb) and lutetium (Lu) [2]. Mortimer and Müller [22] in their chemistry and Meschede [3] in the physics both group the lanthanoids and rare earths synonymously and count 14 elements from Cerium to lutetium, i.e. without lanthanum itself [4, 3, p. 849]. Often the name lanthanides is used instead of lanthanoids. The ending—id however indicates a certain chemical structure like a sulphide whereas the ending—oid means that something is similar, looks or behaves the like. So the word lanthanoids would be the correct one to use as the REE are not derivations

V. Zepf, *Rare Earth Elements*, Springer Theses,
DOI: 10.1007/978-3-642-35458-8_2,

from lanthanum but they are similar to lanthanum. Even though that would preclude lanthanum itself from the group, the inclusion of lanthanum has become common use [2].

2.1.2 The Terms 'Rare' and 'Earth'

The name 'Rare Earths' is misleading as the name itself implies an earthen material which is not ubiquitous. To start with the easy part, the word 'earth' was a common denomination for an oxidic material, usually a metal oxide, i.e. a compound of an element with oxygen. In the German language there is another cause for error as the German word for earth, 'Erden', can be misunderstood with the expression 'Steine und Erden', which in the German language stands for materials from pit and quarry industry. Of course the 'Steine und Erden' have nothing to do with RE.

The more difficult part is the adjective 'rare'. Reiners [5] suggests that 'rare' has more of an etymological background, as 'rare' was used from the late 15th century onwards in the sense of something strange, extraordinary, astonishing [5]. REE are not per se rare and in abundance lists they range somewhere from the second half to the beginning of the last third compared to the other elements. This abundance topic is a research field of the geochemistry where Frank W. Clarke, as chief chemist of the United States Geological Survey founded in 1884, published a first estimate of elemental abundance in 1889. Another major protagonist is V. M. Goldschmidt who decisively invented methods to determine the abundance of elements in minerals at around the 1920s. More details about the initial phases of geochemistry and abundance tables are elaborated by Mason et al. [14]. Initially earth samples were taken and abundances derived thereof. Later, meteorite data were taken as well to determine abundances of the stellar, solar and earth elemental abundances [6–8]. Today there are numerous abundance tables available with usually different approaches, assumptions and geographical sample areas, so that no uniform list can be presented. Rudnick and Gao [9] give a fundamental insight into the composition of the continental crust and the research in the field which they call 'myriad studies on continental crust composition' [9]. They come up with a new suggestion for the bulk composition differentiated in an upper, middle, lower and total crust value. Table 2.1 shows some selected elements and the proposed composition of Rudnick and Gao [9]. As a comparison to illustrate the differences also the abundance estimates of Binder [10] and Allègre et al. [6] have been added.

Table 2.1 shows that there is no uniform estimate: Rudnick and Gao [9] put the least abundant REE, lutetium and thulium, behind platinum and gold. Binder [10] lists lutetium and thulium before silver, gold and platinum, i.e. more abundant whereas in Allègre et al. [6] they again range behind, i.e. less abundant than gold and platinum. So the abundance tables are more or less of theoretical value. The basic data Binder [10] is using in his encyclopedia is not known. Allègre et al. [6]

Table 2.1 Elemental composition of the Earth

Composition of the Earth-comparison of different authors' data and levels of detail						
Element	Upper crust	Middle crust	Lower crust	Total crust	Composition of the Earth's crust	Bulk composition of the Earth
	Rudnick and Gao [9, p. 53f] (μg g^{-1}) (≈ppm)				Binder [10, p.776f]	Allegre et al. ([6], p. 61)
Sc	14	19	31	21.9	20 ppm	10.1 ± 2 ppm
Y	21	20	16	19	31.5 ppm	2.4 ± 0.2 ppm
Cu	28	26	26	27	68 ppm	64.7 ± 5 ppm
La	31	24	8	20	35 ppm	415 ± 10 ppb
Ce	63	53	20	43	68 ppm	1088 ± 20 ppb
Pr	7.1	5.8	2.4	4.9	9.5 ppm	165 ± 5 ppb
Nd	27	25	11	20	40 ppm	814 ± 10 ppb
Sm	4.7	4.6	2.8	3.9	7.5 ppm	259 ± 3 ppb
Dy	3.9	3.8	3.1	3.6	6.2 ppm	424 ± 10 ppb
Tb	0.7	0.7	0.48	0.6	1.2 ppm	66.6 ± 5 ppb
Lu	0.31	0.4	0.25	0.3	0.81 ppm	42.5 ± 2 ppb
Pt	0.5	0.85	2.7	1.5	0.004 ppm	1562 ± 40 ppb
Au	1.5	0.66	1.6	1.3	0.0041 ppm	102 ± 20 ppb

REE marked bold italic

base the determination of abundance tables or the bulk chemical composition of the earth on the comparison of terrestrial samples with meteorite data. As earth samples can only be gained from the very outer parts of the earths' crust, one way to attain knowledge is to use meteorite data which is assumed to constitute the original material composition shortly after the big bang. So the idea is to infer from these chondrites to the earths composition. Research showed that there is a high similarity of elemental composition of a special type of chondrites, the so called CI chondrites, and the earth's composition gained by terrestrial samples.

Numbers for the geosphere are usually given in percent of mass (%); data for the earths' crust is given in parts per million (ppm), respectively grams per metric ton (g/t) [10]. The numbers allow the calculation of theoretical quantitative data but they do not imply that more abundant elements are more easily exploitable as less abundant ones. As not all elements occur in pure or elemental form but usually as one constituent amongst many others within a mineral, it depends on the grade if an element mining or extraction is feasible. Skinner [11] introduced the term 'mineralogical barrier' by which he distinguishes abundant elements where mining is feasible. The barrier also determines the required grade for scarce elements to allow economic exploitation [11]. The Clarke-value is another determination of necessary enrichment grade for economic mining (e.g. [12]). As here also economic mining is addressed, of course this value is dependent on available technical options and commodity prices. So the value actually has to be amended as soon as prices or technology and thus extraction cost change.

The geochemistry tries to determine the petrogenetic processes which led to the minerals that can be found today [13, 14]. Usually endogen processes are assumed when in 2011 Willbold et al. [15] determined that 'iron-loving' metals like gold are surprisingly abundant in the earths' crust. So they came up with a new probable explanation that a terminal bombardment of meteorites after the earths' core formation around 4 billion years ago contributed to the earths' elemental composition. The actual composition respectively abundance is even more difficult to determine and as such the direct economic value of abundance tables is questionable.

In this context another question arises when the theoretical abundance data are compared with common mineral compositions. It could be asked: where are all the 'missing' REE? The abundance tables show a higher or near similar abundance for neodymium than for lanthanum: 35 ppm (La) and 40 ppm (Nd) [10], 32 ppm (La) and 38 ppm (Nd) [16], 415 ppb (La) and 814 ppb (Nd) ([6] or 31 $\mu g\ g^{-1}$ (La) and 27 $\mu g\ g^{-1}$ (Nd) [9]). This means that in general there should be more neodymium available in the earths' crust than lanthanum. This is applicable for the composition of the earth as such, but similar relationships are known for the composition within the geosphere alone [10]. However, looking at the major REE minerals like e.g. the bastnaesite of the Bayan Obo type (explanation in following chapter) the concentration is given as approximately 25 % for La_2O_3 and 17 % for Nd_2O_3 [17] or 33.27 % La_2O_3 and 12.02 % for Nd_2O_3 [18, p. 837]. For monazite, the next important REE mineral there are similar numbers given [18, p. 837]. So where is the missing neodymium? There most probably is no undetected new mineral which hosts all the Nd, moreover the chemical characteristics of the REE and their behavior in melts lead to a better incorporation of lanthanum into mineral genesis than neodymium. Due to a higher ionic radius lanthanum is more readily built into rocks than neodymium. As a consequence neodymium remains more dispersed in surrounding rocks and does not aggregate enough. More substantial information is given by Stosch [19].

Eventually, as a conclusion it has to be noted that the mere position in an abundance table does not at all reflect any actual proof of an economic exploitation. So the message that REE are more abundant in the earth's crust than gold or platinum is of no economic use and rather misleading information.

2.1.3 Light, Medium and Heavy REE

Next, the REE are usually attributed to the groups of the light REE (LREE), the heavy REE (HREE) and sometimes to the medium or middle REE (MREE). The grouping to these three classes is not consistent among different authors, as can be seen in Table 2.2. The only reasonable *argument* or the grouping of the REE into ceric and yttric is given by Röhr [20] and the USGS [21]. They argue that the electron configuration is the basis for the distinction of the two groups.

Table 2.2 Rare Earth elements—overview of selected characteristics

Rare Earth Elements - selected characteristics, chemical data and the grouping into LREE, MREE and HREE

Atomic number	21	39	57	58	59	60	61	62	63	64	65	66	67	68	69	70	71
Element	**Sc**[3]	**Y**[3]	**La**	**Ce**	**Pr**	**Nd**	**Pm**	**Sm**	**Eu**	**Gd**	**Tb**	**Dy**	**Ho**	**Er**	**Tm**	**Yb**	**Lu**
Orbitals[1]	$3d^14s^2$	$4d^15s^2$	$5d^16s^2$	$4f^26s^2$	$4f^36s^2$	$4f^46s^2$	$4f^56s^2$	$4f^66s^2$	$4f^76s^2$	$4f^75d^16s^2$	$4f^96s^2$	$4f^{10}6s^2$	$4f^{11}6s^2$	$4f^{12}6s^2$	$4f^{13}6s^2$	$4f^{14}6s^2$	$4f^{14}5d^16s^2$
Oxidation states[1]	+3	+3	+3	+3 +4	+3 +4	+2 +3 +4	+3	+2 +3	+2 +3	+3	+3 +4	+3 +4	+3	+3	+2 +3	+2 +3	+3
Color[1]			no color	no color	yellow green	violett	pink	yellow	nearly no color	no color	nearly no color	yellow green	yellow	pink	green	no color	no color
Atomic radius [pm][1]	162	180	187	182	182	181	181	180	204	179	178	177	176	175	174	193	174
Ionic radius [pm][1]			103	101	99	98	97	96	95	94	92	91	90	89	88	87	86
Atomic mass [u][3] rounded	45,0	88,9	138,9	140,1	140,9	144,2	144,9	150,4	152,0	157,3	158,9	162,5	164,9	167,3	168,9	173,0	175,0
Mass in geosphere [%][3]	$5 * 10^{-4}$	0,003	0,002	0,004	$5 * 10^{-4}$	0,002	10^{-19}	$6 * 10^{-4}$	10^{-5}	$6 * 10^{-4}$	$9 * 10^{-5}$	$4.2 * 10^{-4}$	10^{-4}	$2 * 10^{-4}$	$2 * 10^{-5}$	$3 * 10^{-4}$	$7 * 10^{-5}$
Boiling point [K][2]	3104	3611	3730	3699	3785	3341	~3000	2064	1870	3539	3396	2835	2968	3136	2220	1466	3668
Melting Point [K][2]	1814	1795	1194	1072	1204	1294	1441	1350	1095	1586	1629	1685	1747	1802	1818	1097	1936
Density [kg/m³ at 293K][2]	2989 [273K]	4469	6145	8240 [298K]	6773	7007	7220	7520	5243	7900 [298K]	8229	8550	8795 [298K]	9066 [298K]	9321	6965	9840 [298K]

Classification of metals	Sc	Y	La	Ce	Pr	Nd	Pm	Sm	Eu	Gd	Tb	Dy	Ho	Er	Tm	Yb	Lu
Classification of metals	Light metals		Heavy metals (Density > 5000 kg/m³)														
Binder [10], p. 342; USGS 2011b (Cordier)			Ceriterden (ceritic earths), LREE								Yttererden (yttric earths), HREE						
Röhr [20]				Ceriterden (ceritic earths), LREE							Yttererden (yttric earths), HREE						
Cesbron [87], p. 5; RÖMPP Online [23][4]		HREE	Light Rare Earth Elements (LREE), Cerium group							Heavy Rare Earth Elements (HREE), Yttrium group							
Pohl [29], p. 212			LREE							HREE							
Kingsnorth [89], p. 4		HREE	LREE					Medium REE (MREE)			HREE						
Stosch [19], p. 2		HREE	LREE				MREE						HREE				
Chen [44]	HREE		LREE							Middle REE					HREE		

Notes and data sources:
(1) Mortimer 2001, p. 501f
(2) Emsley [16], p. 233
(3) Merck 2007. Data for Sc and Y have been extracted from Merck PSE
(4) keyword ‚Seltenerdmetalle'
blank cells - no data was available or no information given

The overview shows - next to some chemical data - the persisting discrepancies in grouping of the REE into the light, medium and heavy fractions.

The electron configurations f^0 (all f-instances are unoccupied), f^7 (half of the f-instances are occupied) and f^{14} (all f-instances are occupied) are considered more stable than the other configurations. So these more stable instances can act as borders for the groups such as to group the elements cerium—gadolinium to the ceritic earths or light rare earth elements (LREE) and terbium—lutetium to the yttric earths or heavy rare earth elements (HREE). A reason for not attributing lanthanum to the LREE is that lanthanum is no f-element, so it does not fit to the LREE-nomenclature. Cerium however can and should be regarded as a 1f-state due to its position within the periodic system of elements and the reactivity of cerium (personal information from Röhr [20]). Stosch [19] gives an explanation for counting yttrium to the HREE. Yttrium stands just above the other REE within the periodic system, so that this gives a hint for similar characteristics within these elements. This is of course true for all REE so that another argument seems more plausible: indeed the ionic radius of yttrium is nearly the same as for dysprosium or holmium and the chemical behavior of yttrium is very similar to holmium so that they can be grouped together into the HREE.

2.1.4 Rare Earths, Rare Elements and Rare Metals

Another misconception results from the similar names: rare earth, rare metals and rare elements. They are not the same and cannot be used synonymously, even though that happens quite often. The REE themselves have been explained already. The group of the *rare metals* does not have a commonly accepted list of metals; instead the arrays differ from author to author. And, over time the allocations of elements to the list change frequently, so that an inherent temporal aspect has to be attributed. Skinner [11] mentioned that the 12 most abundant elements (O, Si, Al, Fe, Ca, Mg, Na, K, Ti, H, Mn, P) in the continental crust account for 99.23 % of the mass of the crust. He therefore introduced all metals with a share of less than 0.1 % on the continental crust as geochemically *scarce* metals [11, p. 563f]. Scarce is often used as a synonym for *rare* thus the definition of rare metals is sometimes derived from Skinner as being the metals with an abundance share of less than the 0.1 %. The group of rare elements now generally addresses all elements, not only metals. Within this nomenclature the rare earths consequently can be considered as a subset of the rare metals which again can be seen as a subset of the group of the rare elements.

2.2 Chemical and Physical Properties

Details about the special chemical and physical properties are explained in appropriate scientific encyclopedias and standard works about chemistry and physics like Mortimer and Müller [22], RÖMPP Online [23], Meschede [3] and many more. The three given books build the basis for this short chapter and will not be cited within the chapter any further. As these books give a very thorough insight into the topic only a brief description is given here about some characteristics which are deemed important to understand the complex creation, extraction and separation processes of the REE.

In the periodic system of the elements the REE are transition metals and belong to the third subgroup, the scandium-group, together with scandium, yttrium and the group of the radioactive actinides. The lanthanoids demonstrate special electron configurations on the atomic level. The atomic cores are surrounded by electrons on several orbitals and with increasing atomic number protons, neutrons and electrons are gained. The electrons usually attach to the outer orbitals. For the lanthanoids however, not the outer shell is filled but a deeper lying orbital accommodates the new electron(s): the 4f-orbital. So the 'outer appearance remains the same for all lanthanoids and as the inner orbitals do not influence the chemistry much, this explains the chemical resemblance of the REE. The physics indeed is not touched by this chemical fact so that the elements show different physical behavior. Characteristic for the lanthanoids is their common oxidation state of +3, whereas some REE show also other states (see Table 2.2). Preferred stable states are attained by empty, half or full occupied orbitals, i.e. La^{3+} (f^0), Gd^{3+} (f^7) and Lu^{3+} (f^{14}). Other stable configurations are Eu^{2+} (f^7) (which serves as a good reduction-agent), Yb^{2+} (f^{14}), Ce^{4+} (f^0) (as a good oxidation agent) and Tb^{4+} (f^7).

A specialty of the electron configuration is that with increasing atomic number the ionic radii get smaller; known as the lanthanide-contraction. This fact results in ionic radii of most trivalent REE similar to the radii of Ca^{2+}, Th^{4+} and U^{4+}. The slightly bigger Eu^{2+} has a similar radius to Sr^{2+}. This coincidence is important for the understanding of petrogenetic processes as most REE can and do replace elements with similar ionic radii. So REE can be found in rocks which contain Ca, Th, U and Sr.

For the physical characteristics, the REE show very diverse behaviors which make them especially useful for a wide range of applications. In magnetism some of the REE like Gd, Dy, Er but also Nd and Sm show complex potentials which can be and are used for magnets manufacturing [3, p. 394]. Some REEs provide sharply defined energy states which can be efficiently used in lighting and laser applications.

In summary there is a wide variety of similar and special characteristics inherent in the REEs. Especially the different atomic structures and states lead to some unique properties. Thus a complex and huge application field results from these properties which show similar but also completely different characteristics. This should give also an idea that REEs cannot be compared and dealt with as one single element.

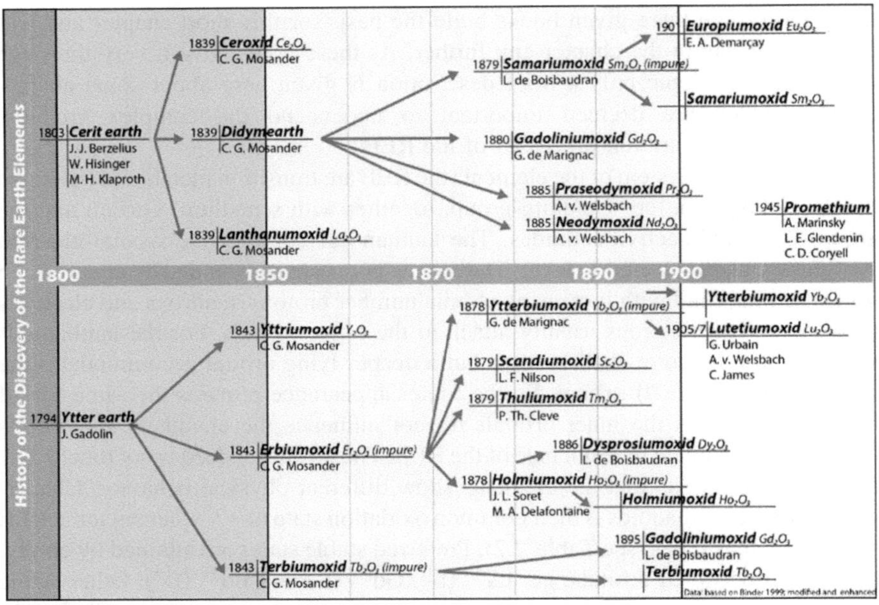

Fig. 2.1 History of the discovery of the REE

2.3 The Long History of the Discovery of the REE

The following story has been put together on basis of von Welsbach [24], Plohn [25], Greinacher [26], Trueb [27], Binder [10] and Brock [28]. They all give interesting insights in the long lasting discoveries of the rare earths (see Fig. 2.1). This is deemed interesting as it shows one of the basic problems: the separation of the individual REEs; which is still a decisive challenge today.

The fascinating story of the rare earths began around 1787 when the Swedish artillery officer Karl Axel Arrhenius, whose hobby was mineralogy, found a dark mineral in a feldspar pit near Ytterby, close to Stockholm. Greinacher [26] tells that the discovery leads back to a mine foreman who found a black mineral in 1788, which initially was called Ytterbite and later Gadolinite. The Finnish chemist Johannes Gadolin (1760–1852) studied this mineral and realized that this new material was of oxidic structure and called it 'rare earth'. In 1794 finally, he discovered within that mineral a new element which was called ytterearth and later named Gadolinite. It turned out that in this mineral nine elements were hidden, but it took around 100 years to discover them all. Around 1803 the German pharmacist Martin Heinrich Klaproth (1743–1817) and the Swedish Jöns Jacob Berzelius (1779–1848) and Wilhelm Hisinger (1766–1852) discovered nearly at the same time but with independent effort another mineral in an abandoned pit near Bastnas, Sweden, which was later called bastnaesite. From that mineral they succeeded in the extraction of an element which upon heating showed a yellowish material. They called it an 'ochroite earth' and later named it cerium, after the 1801 discovered small planet Ceres. Again this ceric earth contained its secrets nearly 100 years until the last ceric earth Europium was discovered [27]. Quite some time after Klaproth and Berzelius discovered the ceritic earth, it happened to the Swedish chemist Carl Gustav Mosander (1797–1858) in 1839, to separate the ceritic earth into the parts cerium (ceroxide), lanthanum (lanthanumoxide) and didym (didym-earth). Cerium often occurs in the so called Mischmetall to which it contributes about 50 %. The name lanthanum leads back to the Greek "lanthanein", which means 'being hidden'. The name didym originates on the Greek word 'didymos', i.e. twin, and was chosen due to the similarity of didym to lanthanum. It then took until 1879 when French Paul-Emile Lecoq de Boisbaudran (1838–1912) extracted samariumoxide from the didym-earth. Samarium was named after the Russian mining-clerk Samarsky, who also discovered the mineral samarskite. In this samarskite Boisbaudran demonstrated by spectroscopy the existence of samarium. In 1880 the chemist and professor Jean-Charles Gallisard de Marignac (1817–1894) from Geneva isolated another element out of the didym-earth, the gadolinium (oxide). The element was named after the chemist Gadolin. A further prominent scientist was the chemist Carl Auer von Welsbach (1858–1929) from Vienna. In 1885 he proved that didym-earth actually consisted of two elements: praseodymium and neodymium. The names were derived from the (Greek) attributes 'praseios' = leek-green (praseodymium—leek-green twin) respectively 'neos' = new (neodymium—new twin). Finally, in 1901 the French chemist Eugène Anatole Demarçay (1852–1903) succeeded in the discovery of the last ceritic earth, the europiumoxide. He extracted it out of the samariumoxide, which was thought to be chemically pure until that time. Named was europium after the continent Europe.

The ytter-earths as well can exhibit a similar history. After Gadolin presented the ytterearth in 1794, it took until 1842/1843 when Mosander extracted out of the ytterearth the ultimate yttrium, impure erbiumoxide and impure terbiumoxide. Mosander just shortly before discovered the cerium oxide and didym earth.

After another 30 years, in 1878, de Marignac could separate ytterbiumoxide out of that impure erbiumoxide. In 1879 further oxides could be extracted: scandiumoxide by the Swedish Lars Nilson (1840–1899), thuliumoxide by the Swedish Per Theodor Cleve (1840–1905) and (impure) holmiumoxide by the Geneva scientists Marc Delafontaine (1837–1911) and Jacques-Louis Soret (1827–1890). Scandium was named in honor of Scandinavia, thulium after the Nordic mythical island Thule and holmium after the Latin name of the Swedish capital Holmia. In the impure holmiumoxide de Boisbaudran found in 1886 the dysprosiumoxide. Its name was derived from the Greek word 'dys', which stands for 'strenuous attainable'. The next to last discovery was in 1905 the separation of lutetiumoxide from the impure ytterbiumoxide which was demonstrated by three scientists independently from another: Auer von Welsbach, Georges Urbain and Charles James. The name goes back to the Latin name of the Paris, Lutetia. Finally, in 1945 in the USA, the radioactive element promethium was chemically isolated. This discovery was ascribed to both Jacob A. Marinsky, Lawrence E. Glendenin und Charles D. Coryell. Named was promethium after the titan Prometheus out of the Greek mythology. He robbed the heavenly fire for the humans. Other scientists were not as successful as the ones described. One of the *unlucky* scientists was William Crookes who tried to identify the individual REE by means of the—at that time new—radiant spectroscopy. It turned out however, that impurities in the mineral prevented him in participating in the discovery of the REE during the Rare Earths Crusade [28].

Eventually it took more than 150 years to identify and discover all 17 rare earth elements. The first important use and economic application goes back to Auer von Welsbach. He was, next to being a brilliant scientist, also interested in the commercial use of his findings. First von Welsbach soaked woolen threads with salts of the rare earths and put the threads to the flame of the Bunsen burner and realized that this was the first real light source, as until that time light was more a byproduct of heating. Such the Auerlight was invented. Von Welsbach managed, by use of fractionized crystallization, to separate the individual rare earths. Initially he used lanthanum, later primarily thorium oxide and about 1 % cerium for the lamps. The main raw material was ballast, monazite sand, of Brazilian ships which was abundantly available so that cerium was there in excess. Then von Welsbach melted cerium with iron and thus discovered the Auermetal, a prerequisite for lighter flints, which are still used today. At that time also the functional use of Ce for glass production and UV protection was discovered [25]. So the first commercial applications for REE were the Auerlight, flintstones and the use of RE fluoride as wicks in arc light carbons [26].

2.4 Geology and Geochemistry of the REE

At first a short explanation of geological and mineralogical terms seems appropriate, as these definitions are usually unfamiliar to non-mineralogists and non-geologists. Knowing these backgrounds can contribute to minimize the inaccuracies around the availability of the REE.

A mineral is a chemically homogenous, naturally formed part of the earth. With a few exceptions minerals are inorganic, solid and crystalline. A crystal is a solid, homogenous and anisotropic body with a three-dimensional periodic array of the components, i.e. atoms, ions or molecules. Here the elements are addressed. Rocks are comprised out of mineral aggregates and form vast geological bodies. Ores are rocks or mineral aggregates which contain metals or metallic compounds. A spatially confined area of concentrated minerals is called occurrence and if it is exploitable the words *deposit*, *ore deposit* or *mineral deposit* are used. Specific details are given in mineralogical literature (e.g. [12, 14, 19, 29, 30]).

The basic principles of petrogenesis are known to a certain degree today so that potential REE aggregations, occurrences and deposits can be judged reasonably well. The geochemical behavior of the REE and above all the ionic radii are responsible for the enrichment of REEs. The ionic radii of common rock forming elements like aluminum, chromium, iron or sulfur are smaller than the ones of the REE, so that the REE cannot be built into the crystal structure of general rocks. So they are no major rock forming elements, instead they aggregate in residual fluids and enrich there as accessories [29, 31].

All REE are lithophile elements, i.e. they occur together with oxygen usually in oxide, silicate or phosphate combinations. As well, REE never occur in pure, i.e. native form but always together with other REE as accessories in minerals. There is mainly an aggregation as LREE or HREE and some complex mineral ores [23]. Typical REE minerals are monazite and bastnaesite with high grades of LREE or ceritic earths and xenotime which contains higher grades of HREE or ytter earths. Further REE containing materials are the ion-adsorbing clays, which are residues from alteration. Today several hundred REE containing minerals are known, however only very few have been identified to be of economic importance.

In general rare elements can enrich under magmatic processes. Often the REE enrichment is tied to partial anatexis (melting of parts of the earths' crust) and the subsequent magmatic differentiation, i.e. fractionized crystallization. The elements aggregate during several steps in fractions of similar or compatible shape. The complex processes shall not be explained here as they are well explained in respective mineralogical and geochemical literature as stated above.

Typical occurrences of REE are in granites, pegmatites, carbonatites and perovskites [12]. Granites together with other rocks represent the most important group of plutons (magmatic rocks which crystallized below the surface). Pegmatites are coarse to huge grained magmatic rocks which can contain single crystals with sizes of several meters. In principle plutons can be shaped as pegmatites. Carbonatites are of intrusive character and thus connected to volcanic active intracontinental rift-zones but also hot-spot-areas. They are usually enriched with phosphate-minerals like apatite and monazite, but also bastnaesite and others. Perovskites are oxidic combinations of a metal and oxygen, in this case $CaTiO_3$, whereas Ca can be replaced by REE. Perovskites occur as accessories in Carbonatites and some other magmatic formations [12]. A look at the present REE deposits reveals the just mentioned geological formations. Consequently, potential future mines are found in similar rock formations. Further details are given by Okrusch and Matthes [12], Stosch [19] or Mason et al. [14]. Several authors provide lists with REE containing minerals and deposits, e.g. Kanazawa and Kamitani [32]; Jones et al. [33] and Orris and Grauch [34]. Especially Jones et al. [33] give a vast list of about 200 known REE containing minerals. Elsner [35] states, that monazite and bastnaesite are the major minerals worth mining. Both are containing mainly LREE. Another relatively abundant mineral is xenotime which contains HREE, specifically yttrium, but also dysprosium, holmium, erbium, thulium, ytterbium and lutetium [35, p. 92f]. Furthermore the RE containing clays or ion-adsorbing clays, which occur mainly in southeast China, shall not be forgotten as they provide the majority of HREE nowadays.

Monazite, usually shown as [(Ce, La, Y, Th) PO_4] [29], sometimes a bit simpler as [Ce (PO_4)], can also incorporate Nd and Th as these can substitute Ce within the monazite [12, p. 116]. So either way the possible accessory element Th represents the *problem* as Th is radioactive and requires special handling. During mining and processing the material has to be handled according to specific laws and regulations. And not to forget, the tailings have to be treated in accordance to the regulations as well, what poses a big challenge. This does not mean that monazite cannot be mined, but the handling associated with radioactive materials handling is intensive both in time and cost. At the beginning of the 20th century REE mining was predominantly from monazite so that the first era of REE mining is called the *monazite-era* [36]. Monazite is primarily found in granites, pegmatites and carbonatites. Monazites secondary occurrence is as placer deposits, marine and beach sands. Together with rutile, ilmenite and zirconium, monazite occurs as heavy sand deposits and was mined around the 1950s in several countries, e.g. as beach sands in Florida, USA or placer alluviums in Idaho, USA (Bureau of Mines, [37], p. 1178), in Australia, India or Brazil. The heavy sands mining activities have been reduced, mainly due to the radioactive materials handling challenge.

Bastnaesite, a fluorcarbonate mineral, is the other important mineral in REE mining. The basic chemical formula shows [(Ce, La) (CO_3) F] whereas varieties exist with different balances of La and Ce. Next to the given formula also other LREE can be contained like Pr, La but also Y (e.g. [31, 33]). Bastnaesite is the main mineral that was exploited in the Mountain Pass Mine, California, USA. This mine gave name to the era following the monazite-era: the *Mountain Pass Era*.

Xenotime, [(Y, Yb) (PO_4)], another phosphate contains higher grades of HREE, especially Roskill [17] states that the four mentioned minerals, monazite, bastnaesite, xenotime and the lateritic clays account for about 90 % of the economic production. Table 2.3 details in more depth the typical elemental contents of the minerals mined at present, respectively in near future (the Mt. Weld churchite).

Table 2.3 shows various deposits and their typical RE contents. It can be seen, that monazite and bastnaesite provide high percentages of the LREE whereas the HREE only occur in very small percentages, sometimes only in traces. The xenotimes and lateritic ion adsorption clays show a relative diverse spectrum with a clear emphasis on HREE. Roskill [17] provides slightly different data, which they extracted from NDRC, metal-pages and direct contacts with RE processors. It has to be noted eventually that these contents data represent average values. The mineral compositions may and obviously do differ from ore body to ore body and sometimes even within one ore body; variations and changes are common.

In 2011 a team of Japanese scientists discovered a new geological setting as a new potential resource for REE in the Pacific Ocean. They measured more than 2,000 seafloor samples and found high concentrations of REE and Yttrium within some of the samples. From these samples estimates were derived which show huge REE quantities available. The concentrations and mineral compositions show promising structures which would allow relatively easy refining. However, the team clearly stated that the samples were taken from depths mostly between 4,000 and 5,000 m, and they as well estimate that an extraction *today* is not possible. Eventually the discovery then shows a potential for mid to far term future extraction [38]. Indeed the occurrence of REE in deep-water polymetallic nodules is known long since and is investigated in several projects in these days (e.g. ISA,

Table 2.3 Elemental composition of selected REE minerals

Deposit	Mt Weld CLD	Mt Weld Duncan	Mountain pass, CA	Bayan Öbo, Inner Mongolia	Guangdong	Xunwu, Jiangxi	Longnan, Jiangxi
Country	Australia	Australia	USA	China	China	China	China
RE mineral	Secondary monazite	Churchite	Bastnaesite	Bastnaesite	Xenotime	Laterite	Laterite
Lanthanum	25.57	23.93	33.2	23	1.2	43.4	1.82
Cerium	46.9	39.42	49.1	50	3	2.4	0.4
Praseodymium	4.92	4.85	4.34	6.2	0.6	9	0.7
Neodymium	16.87	18.08	12	18.5	3.5	31.7	3
Samarium	2.29	2.87	0.8	0.8	2.2	3.9	2.8
Europium	0.49	0.77	0.1	0.2	0.2	0.5	0.1
Gadolinium	1.33	2.15	0.2	0.7	5	3	6.9
Terbium	0.13	0.29	Trace	0.1	1.2	Trace	1.3
Dysprosium	0.31	1.36	Trace	0.1	9.1	Trace	6.7
Holmium	0.04	0.21	Trace	Trace	2.6	Trace	1.6
Erbium	0.113	0.46	Trace	Trace	5.6	Trace	4.9
Thulium	0.01	0.04	Trace	Trace	1.3	Trace	0.7
Ytterbium	0.05	0.2	Trace	Trace	6	0.3	2.5
Lutetium	0.02	0.03	Trace	Trace	1.8	0.1	0.4
Yttrium	0.95	5.36	0.1	Trace	59.3	8	65
Checksum	99.993	100.02	99.84	99.6	102.6	102.3	98.82

Notes and sources

Lynas [87] Roskill international Rare Earths conference presentation. 16. Nov. 2011

USGS and Hedrick [88] Minerals yearbook 2000. Rare Earths. Hed rick assembled a comprehensive list of deposits and their RE contents. For each deposit further data sources are given

Data given in (% REO)

w.y.) Here the already known static range of 850 years (on land) has to be set as a contrast. There is no physical or mineral shortage, but rather a technological and probably economical issue to be solved. So the usefulness of deep-water mining has definitely to be seen in this context.

2.5 Deposits, Current and Potential Future Mining

In summary of the chemistry and geology of the REE it can be said that the petrogenesis of the REE is reasonably well known. In turn potential and possible occurrences can be derived and are more or less well known since a while already. There is also evidence of deposits in several places of the earth. Gupta and Krishnamurthy [39], Orris and Grauch [34], Jackson and Christiansen [40] or Berger et al. [41] give comprehensive information and lists of known mines, deposits and occurrences. So even before the increase of interest in REE since about 2008, it was known that from a geological point of view there is no physical shortage of REE. The USGS in 2011 announced the REE reserves with 110 Mio t which equates to a static range of approximately 850 years [42]. Still, that value seems to be too low as today around 180 companies work on about 270 projects around REE prospection, exploration and mining [43]. Not all of these new prospection and exploration data are incorporated yet in the data given by the USGS. Looking at these facts, a strange contradiction becomes apparent. In principle there should not be a shortage situation. But it does actually exist! At least that is, what the media tell us. So the question again is how and why this can be or asked the other way around, what is actually the case? Is there a real shortage or is it only an anticipated shortage? And if a real shortage situation should exist and there is a range of 850 years, what good for is the information of 850 years of range? If nothing happens, this static range is of no practical use at all. Usually the monopolistic position of China is put into the field, which could or would prohibit new mining ventures. Of course this scenario is possible, but the question remains whether that is a reasonable argument for not pursuing any steps but pointing fingers at China instead? These questions should raise the awareness, that there is more than just one problem involved. Nevertheless the facts contain potential solutions. Having this in mind the status quo of REE mining and deposits will be illustrated briefly together with the description of some projects which are said to start production within the next one to three years. The ordering has been done by country and alphabetically.

2.5.1 China

The three major mining areas in China are in Baotou, Sichuan and Jiangxi which together host 88 % of the Chinese deposits. Affiliated to Baotou is the Bayan Obo mine in Inner Mongolia where 83 % of the Chinese deposits are concentrated. In

Sichuan province as well as in the southern deposits 3 % each are located. The remaining 11 % are dispersed throughout China [44]. The largest mining site is situated in Bayan Obo, Inner Mongolia, which belongs to the Baotou Iron and Steel Corporation. The huge polymetallic REE-Fe-Nb deposit is exploited with iron being the primary product and REE and Nb being secondary or by-products. The principal REE minerals in Bayan Obo are bastnaesite and monazite with major Ce, La and Nd contents [45].

In Bayan Obo REE mining began in the late 1950s, even though the deposit was discovered already in the late 1920s. Today the REE are mined as a by-product of iron ore mining. More detailed information is provided by Hurst [1] and the archives of the Bureau of Mines (today: USGS), 'Minerals yearbook metals and minerals (except fuels)' in the chapters 'Minor metals' and later in own chapters about 'Rare earths'.

Wu et al. [46] show that the elemental distribution of the Bayan Obo bastnaesite and monazite are very similar and comprise about 26 % La_2O_3, 50 % Ce_2O_3, 5 % Pr_2O_3, 16 % Nd_2O_3, 1 % Sm_2O_3 and the remaining REE less than 1 % each.

A further important formation is situated in Mianning, Sichuan, called the *Mianning-Dechang Himalayan REE Metallogenetic Belt* along the Yangtze Craton. Scientists discovered five potential REE deposits in the carbonatite [47, 48]. The Sichuan mines exploit bastnaesite with about 37 % La_2O_3, 47 % Ce_2O_3, 4 % Pr_2O_3, 10 % Nd_2O_3 and all others below 1 %. An interesting side note is that Sichuan lies in a potential earthquake zone. The strong 2008 Sichuan earthquake occurred along the Longmen Shan fault which is close to the mineral belt. Mining projects thus have to keep in mind potential additional dangers from tectonics.

In southeastern China REE containing lateritic clays are known. In several provinces around Jiangxi these ion-adsorbing clays are exploited which show very diverse distributions ranging from 2 to 30 % of La_2O_3, 1–7 % Ce_2O_3 and Pr_2O_3, 3–30 % Nd_2O_3 and 2–7 % Dy_2O_3 [46, table p. 291]. These southern provinces are featured as middle to heavy REE deposits where mainly Gd to Lu, Sc and Y are mined [44]. As especially Dy is required for improving the temperature stability of NdFeB magnets, these deposits are of great importance. However, there are concerns about these deposits as "…their ion absorption [sic!] (laterite) clay deposits are expected to be depleted by 2025 or earlier. This leaves them [China] with only one known xenotime deposit in Guangdong as their primary source of heavy rare earths". [49, p. 6].

Further occurrences of REE have been discovered in weathering crusts of basalt in Yunnan, Guizhou and Sichuan provinces which are subject of further research for potential new deposits (Yang et al. [50]). Occurrences and deposits are also probable in the Tsagaan region in Mongolia and China [51]. More research information is given e.g. in the Journal of Rare Earths.

2.5.2 India

In India the production of REE products is under the control of the Indian Rare Earths Limited (IREL), a governmental undertaking under the administrative control of the department of Atomic Energy. It was founded in 1950 with the primary intent to produce monazite in a commercial scale for the recovery of thorium. Today IREL operates four units, the Orissa Sand complex (OSCOM), the Chavara Mineral Division, the Manavalakurichi (MK) Mineral Division—all exploiting mineral sand deposits, and the Rare Earths Division (RED) in Aluva where a chemical plant processes monazite produced by MK. It is stated, that only MK produces 3,000 t of monazite per year and that RED has a production capacity of 3,600 t of monazite. REE products, Ce, Nd, Pr and Sm are offered in different grades and states. The major product however does not seem to be REE, instead the emphasis is on thorium and uranium in mantle grade quality and the export of titanium oxide as pigment [52]. The USGS usually lists India as a producer of REE with an annual production of 2,700 t [42]. This value hasn't change since years. Both data sets, the ones from IREL and USGS, cannot be proofed for reliability and validity.

2.5.3 Brazil

In Brazil heavy sands mining is known since the turn of the 20th century. Several mining areas which exploit heavy and monazite sands are dispersed throughout Brazil with five areas encompassing REE rich sands, including also thorium and uranium: Buena, Cumuruxatiba, Guarapari, Mato Preto and Sapucai all located in the southern and southeast regions. The beginning of the monazite mining can be traced back to the invention of the Auer-Light which required cerium and thorium for production. As well thorium was used in therapeutic applications so that the thorium refining industry experienced a boom during the first decades of the twentieth century [53, p. 190f, 54]. Today there is a problematic legacy associated with the mining of monazite and the naturally occurring radioactive materials (NORM). As the tailings have not been treated cautiously enough, there are several contaminated areas which require repeated site remediation, as previous works have not been accomplished accordingly. Today Brazil tries to improve future actions and NORM handling based on the lessons learned from previous monazite processing [55]. The present monazite mining and processing is under the head of the national atomic agency, the Indústrias Nucleares do Brasil (INB) [56]. In 2005 and 2006 the INB produced 958 t of REE out of monazite heavy sands, without further detailing the elemental share [57]. The USGS gives no production data for Brazil in 2005 and 2006 so that even when taking into account that the USGS states 400 t for 'Other countries' in 2006 [58] that does not match and cannot reflect the data given by Rezende [57]. This inconsistency in data again shows the persisting informational dilemma.

2.5.4 Russia, Commonwealth of Independent States and Kazakhstan

Reliable data about Russian production could not be derived. Several sources as mentioned at the beginning of this paragraph state present production or at least potential deposits in Russia and the CIS. The most promising being the Afrikanda complex, Kola peninsula [59], the Khibina and Lovozero alkaline massifs on the Kola peninsula [60] and the huge Tomtor massif in eastern Siberia [41, 61]. For this Tomtor deposit an exact and reliable geographical position could not be determined, rather three possible locations are likely, one being near the city of Yakutsk, the next near the city of Oymyakon, attributed the coldest city of the world, and Berger et al. [41] suggest a position about 1,100 km northwest of Yakutsk. Nevertheless all of the three positions have in common that they are on permafrost soil and they all lack transportation and energetic infrastructure. If this deposit would be exploited the climate, energetics and logistics would pose a big challenge.

Further deposits are supposed to be in Kutessay, Kyrgystan, however no reliable data are available [21].

2.5.5 Australia

In Australia several deposits are known. Initially REE were extracted from beach or heavy sands as a byproduct of titanium mining. The most promising deposit today however is the Mt. Weld mine in Western Australia owned by Lynas Corp. Ltd. Extraction is supposed to start as soon as the Lynas Advanced Materials Plant (LAMP), situated in Malaysia, gets into full commission. This processing facility was the bone of contention as the population raised concerns about the radioactive materials (i.e. thorium) handling associated with the plant. Finally the International Atomic Energy Agency (IAEA) issued a file and reported that they had not found any non-compliance to any IAEA safety standards, however 11 recommendations were issued [62]. The Malaysian government followed these recommendations (Malaysia MTI [63]) which were fulfilled so that the Board announced its approval for a temporary operating license on Feb 2, 2012 [62]. The plan is to ramp up production to 22,000 t REO annually in two steps.

Further projects are the Dubbo Zirconia project by Alkane Resources Ltd., the Nolans project by Arafura Resources Ltd. next to some smaller other ones.

2.5.6 USA

The REE mining in the US began as monazite or heavy sands mining when REE mining was more or less a byproduct of thorium and uranium mining [64]. Major mining activities were in Florida and Idaho where beach sands and placer alluviums respectively were mined. Furthermore monazite sands were known in South and North Carolina, in Lemhi County, Idaho; Wyoming, Alaska and Montana [37, 65]. Exact production data were usually kept secret as the material was needed for the atomic and nuclear energy industries. The radioactive material handling accounted for much of the production price of the raw materials [64, p. 1387]. This additional effort and cost was the main reason for the eventual abandoning of most of the former heavy sands mining activities. It is interesting to note that the names of these old acquaintances appear again nowadays as promising potential new deposits: Lemhi Pass, Idaho and Montana; Diamond Creek, Idaho; Bear Lodge, Wyoming, and Bokan Mountain, Alaska [66, 67].

In early 1949 bastnaesite was found in the Mountain Pass district [68] when subsequent geological mapping showed mineable deposits within the Sulphide Queen Carbonate body which proved as the worlds' greatest concentration of REE known at that time [69]. In 1952 the rare earth open pit mining began under the head of the Molybdenum Corporation of America which was later renamed into Molycorp. The facility also did the beneficiation and processing to provide several REE products. The invention of the color TV in the mid-1960s required considerable amounts of Eu which pushed production [70] and Mountain Pass became the primary REE production facility in the world while the importance of monazite mining diminished. In 1985 ground water samples brought evidence of leaking evaporation ponds due to mechanical failures. Despite remediation activities further contaminations have been identified which contributed arguments to the shutdown of the mining and production facility on Mountain Pass eventually in 1998/2002 [70, 71]. The other argument for the closure at that time was the strong position of Chinese mining projects, which offered REE products cheaper than the U.S. did. In the course of the action of increased focus and demand for REE in 2004 Molycorp finally attained a new 30 year mine plan permit. In 2007 the extraction from tailings was restarted. According to the company's homepage, a full up operation is planned to effectively start in late 2012 with a scheduled production capacity of 40,000 t of REO annually. The idea incorporates the Phoenix Project which is to modernize the installations, and basically to offer an efficient "mine-to-magnet" supply chain. For this, the company attained shares in several downstream businesses like production facilities, magnet producers, wind energy and recycling of REE.

2.5.7 South Africa/Canada

In South Africa the former producing Steenkampskraal mine is planned to reopen production in 2013, according to Great Western Minerals Group [72]. The mine was in operation from 1952 until 1963 when monazite was extracted from underground mining with a subsequent processing in the United States. GWMG, the owner of the mine today, states the aim to become a fully integrated rare earth producer. The company therefore invested in a chain of downstream businesses like exploration projects in Canada, an rare earth alloy processing facility in the UK, and a joint venture with a Chinese company to install a separation facility near the mine at Steenkampskraal [73]. A further South African project is the Zandkopsdrift site [74].

2.5.8 Others

Hatch [75] assembles and updates an 'Advanced Rare-Earth Projects Index' where he lists these projects which fulfill certain criteria concerning resource and reserve numbers and basically the knowledge and capability to actually start a mining project. Out of nearly 400 projects known in mid-2011, twenty projects fulfill these criteria. Next to the projects mentioned above, there are the following enterprises included: Eco Ridge, Eldor, Hoidas Lake, Nechalacho (Thor Lake), Strange Lake and Zeus (Kipawa), Canada; Kangankunde, Malawi; Kvanefjeld and Sarfatoq, Greenland and Nora Karr, Sweden. The German BGR lists several further projects foreseen as exploitable deposits located in southern Africa, Canada, the US, Argentina, Madagascar, Vietnam and Indonesia. Occurrences are even more widespread and including areas in Brazil, Saudi Arabia, Europe and Mongolia (Steinbach et al. [43]).

In summary there are several promising projects in quite advanced stages of development. In how far these projects actually will be realized is not known and will not be commented. Next to the already mentioned projects many others are pursued, but it cannot be verified, how far these projects actually have developed. Several reports are provided by financial analysts and consultants who try to suggest successful investments. This intent often lacks scientific clarity and reliability so that they are not evaluated in this work. As for Hatch [75] and the BGR it can be assumed, that they consider projects which fulfill commonly accepted and expected standards. These are primarily the JORC, the Australasian Code for Reporting of exploration results, Mineral Resources and Ore Reserves. "The JORC Code provides minimum standards for public reporting to ensure that investors and their advisers have all the information they would reasonably require for forming a reliable opinion on the results and estimates being reported" [76].

One more important argument when talking about new mining projects is the required timeframe for opening up a new mine. The development of a mineral exploitation is roughly divided in five to six phases. Starting with the *Mineral Resource Assessment* there follow the *Mineral Exploration Phase,* the *Mineral Deposit Appraisal and the Mine Complex Development*. Until this phase about 5–10 years' time are usually necessary. The required timeframe depends, next to the geological-mineralogical issue, on the jurisdictional facets like ownerships and licensing. Then the *Mine Production* as such exploits the raw material and finally the last phase of *Reclamation and Environmental Restoration* ends the process. Details are well explained in USGS [77], NRCAN [78] and NRCAN [79].

To conclude this chapter it should have become clear, that there indeed are several deposits which could and will be exploited; nearly all of them are outside of China. These could provide enough raw materials to satisfy demand. So it is not a physical scarcity problem, but more a lack of intent, will and probably funding to start new projects. Maybe also the HREE deposits are still not very promising. Details about estimated and projected quantities are described e.g. by Liedtke and Elsner [80] or Hatch [74]. But even when a new project starts, that does not yet mean, that all problems are solved and production does flow; as the example of Lynas and their separation facility shows. Especially the processing and separation of the REE is the real challenge and the environmentally relevant problem; not the mining activity itself. Finally, one fear or possibility remains, that China could change prices of REE on purpose so that potential new mining ventures would become non-profitable from the very beginning onwards. Here however, China issued that it is not the intent to misuse its monopolistic situation as China expects a growing domestic demand so that they are dependent on imports of REE in a midterm timeframe [44].

2.6 Rare Earth Processing

The RE processing is a very important but also complex issue and not very often described in reports. The importance is as such, as for the separation into the individual REE intensive operations are required both as physical and chemical treatment to get the REE. The used chemicals as well as the tailings bear potential hazards to the environment. The processes will only briefly described as e.g. Gupta and Krishnamurthy [39] give some very in depth insights into the various common processes. As well some historical insight is given into the Chinese hydrometallurgy technology in the Journal of Rare Earths [81].

In principle the steps for REE production can be divided in four phases: first there is the mining process as such where REE containing ores are usually mined from open pit projects. Second is the physical beneficiation during which REE concentrates are gained. Then the chemical treatment extracts oxide products which are subsequently separated using extractive metallurgy to get the individual REE.

After the ores are broken or the heavy sands are mined, the raw materials are treated to gain concentrates. These then are chemically treated using acid and alkali solutions. After this initial chemical treatment, the separation is initiated by a variety of processes:

> The various processes for separating individual rare earths from naturally occurring rare earth mixtures essentially utilize the small differences in basicity resulting from decrease in ionic radius from lanthanum to lutetium. The basicity differences influence the solubility of salts, the hydrolysis of ions, and the formation of complex species, and these properties from the basis of separation procedures by fractional crystallization, fractional precipitation, ion exchange, and solvent extraction [39, p. 158].

As some REE appear, next to their trivalent state, also in further states this feature can be used to extract the REE by successive selective oxidation and reduction processes.

The *fractionized crystallization* process is suited for the 'lanthanum end' where the ionic radii differences are the biggest. For Tm and Ho however the process is extremely intense and time consuming. According to Gupta and Krishnamurthy [39], it takes about 15,000 recrystallization processes to obtain close to pure Tm and about 4 years of repeated crystallization processes to extract Ho from rare earth bromate. For this process several salts and double salts are required. The *Ion Exchange* method is a commercially used method.

> The fact that it is feasible to separate a mixture of 15 rare earths into components exceeding 99.99 % purity in one pass through the system and that the process can be scaled up to multi-tonne quantities, with possibilities of recovering and recycling water, and retaining ion and complexant, has led to the continued commercial use of ion exchange even after the entry of solvent extraction in the field. A major limitation… is that the technique is inherently a batch process [39, p. 168].

Hence the Solvent Extraction is a favorable method with the advantage of compact equipment. The process itself is relatively fast, continuous and is suitable for handling of large quantities. The Ion Exchange method is still considered superior in the production of extremely pure materials [39, p. 193].

So in general there are methods known on how to separate the REE. Yet there remains the problem of environmental damage by processing REE, i.e. the handling of chemicals before, during and mainly after the processing. This poses the real threat to the environment next to the radioactive materials storage of some tailings. Basically all known mishaps in the REE mining and processing were due to the toxication of the environment due to spilled chemicals.

2.7 Applications for REE

The applications for the REE cover a huge field as all 17 elements have to be considered. Despite their chemical similar characteristics they have diverging applications. First a list of the major REE uses and applications is given.

2.7.1 Applications of the Individual REE

A list of applications attributed to the individual REE is attached in Annex D. The range of applications within one element changed over time and got more diverse as innovation leapt forward. In the case of neodymium for example the element initially was used for coloring glass whereas today the major use is for the manufacturing of permanent magnets. So around the beginning of the 20th century all neodymium was used for and within the glass industry; today that share became modest and the major share is for magnets production. As this principle is applicable to all individual REE, the list is long and it should become apparent again that a simple grouping of all REE together does not address the diversity of the 17 elements adequately. Either the single REE should be considered individually or at least as part of an application field like the (permanent) magnet production or the glass industry. But this renders a qualitative and quantitative determination difficult as obviously one element can be part of several application fields. Scientific literature gives some extremely detailed insight into various research and scientific fields, especially the book series 'Handbook on the physics and chemistry of rare earths' edited from the beginning in 1978–2010 by Gschneidner and others, since 2011 edited by Bünzli and Pecharsky (Gschneidner and Eyring [82, 83]). With less detail but with more emphasis on the whole life cycle of rare earths, e.g. Gupta and Krishnamurthy [39] and Delfrey [84] cover some REE in quite good detail.

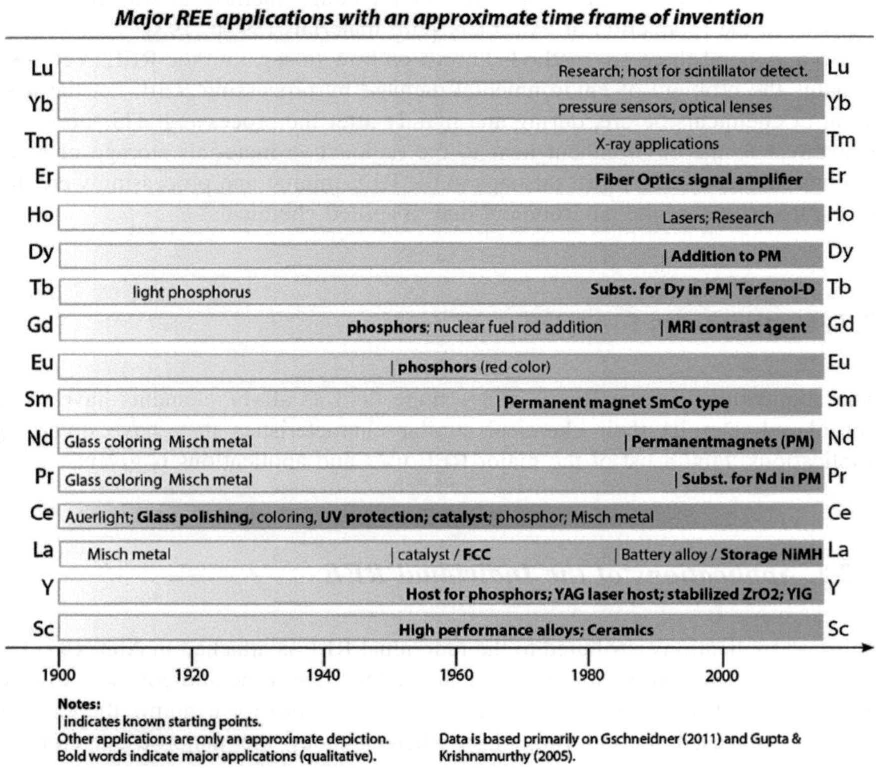

Fig. 2.2 Major REE applications in a historical context

2.7.2 Major REE Applications in a Historical Context

Figure 2.2 shows the major REE applications by element together with a rough depiction on the year of invention for commercial use.

The application—time depictions in Fig. 2.2 do not represent hard dates but rather give a general idea of the beginning of specific uses in about 5–10 year frames. For an overview this depiction is deemed accurate enough; which does not preclude further in depth analysis of selected single cases of invention.

2.7.3 Applications of REE According Functional Uses

Another listing or grouping of REE along application fields or functional use is regularly seen at Chegwidden and Kingsnorth [85] and Lynas [86]. They use the following fields: *magnets*, *metallurgy*, *battery alloys* (sometimes included in metallurgy), *ceramics*, *polishing*, *glass*, *phosphors*, *catalyst* and *others*. A brief arrangement of these applications and the REE used therein is given in Fig. 2.3.

This graph gives an overview of the REE needed in respective functional areas. It would be important to state as well the relative share of the REE to the functional fields. This data set is so far not available and is also subject for further research. This would give a better idea of the real critical areas and potential competitive applications so that appropriate actions could be initiated.

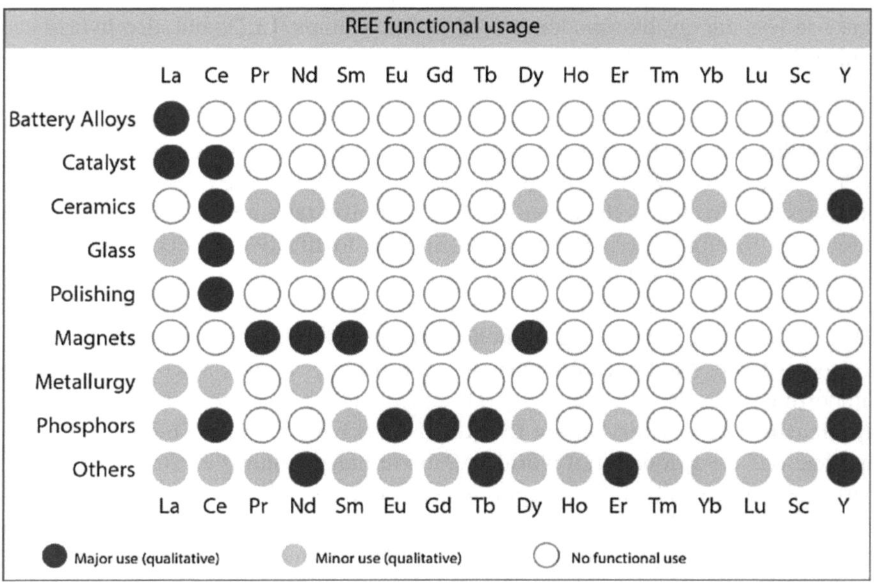

Fig. 2.3 REE and their respective functional use

Applications containing REEs

Area	***(Energy) Efficiency***	***Environmental Protection***	***Digital Technology***	***Medical Applications***	***Military Applications***	***Others***
Objective	reduced consumption	reduced emissions	lifestyle & communication	improved therapy	mobility & reduced weight	better life
Product examples	• Lighting, LED • FCC • Permanent magnets - smaller size products • e-mobility - electro motor • Alloys - reduced car weight • Glass - color & UV protection	• Wind Turbines (Direct Drive REE magnet based) • Car catalytic converters • Diesel fuel additives	• Flat Panel Displays (FPD) - glass, LED • Hard Disk Drives (HDD) • Mobile communication systems • Fiber optics, signal amplifier	• Magnetic Resonance Imaging (MRI) • X-ray systems • Medical additives and contrast agents • Lasers	• electro motors for propellant systems, sensors, guidance systems • special materials (Terfenol-D) • Energy storage (Batteries) • Electro-Motors	• Y x G - Lasers • (super)alloys • Superconductors • Neutronabsorbers • Algae control • Water treatment • Magnetic refrigeration

Fig. 2.4 Applications containing REEs

2.7.4 Applications of REE According Present Discussions About Climate Relevant Uses

A last arrangement is getting more common and was used by Lynas [86] and amended by the author as shown in Fig. 2.4. Here the emphasis is along the present focus on climate relevant issues and uses.

The sorting in Fig. 2.4 is as follows: **energy efficiency** incorporates systems that use less energy like modern energy saving lamps, LEDs but also hybrid cars with e-Motors that reduce gasoline use. Special alloys allow a reduced vehicle weight and thus contribute to less fuel consumption. Finally glass with respective UV protection can help to reduce energy use in buildings. The **environmental protection** block strives for reduced consumption but in a more direct way, i.e. to reduce emissions. WTG reduce the need for carbon based power plants, and catalysts help to clean exhaust emission. The **digital technology** definitely is not in direct ties to climate issues but a contribution to lifestyle and communications. One could argue that this technology allows doing video conferencing and thus saving some travels; if that really is the case is another issue. Nevertheless do some REE allow the manufacturing of modern gadgets which are deemed indispensable for today's life. **Medical applications** are primarily seen as MRI systems, X-ray supporting materials, contrast agents, medical and surgical lasers. **Military applications** have a very strong focus in the US but can mainly be reduced to the applications used in the civil world. I.e. the military requires electro motors using permanent magnets both for weapon systems and expendables like bombs, rockets, etc. They also need computers, communication gadgets, special lightweight steels, glass for several applications or storage systems. There are some materials like Terfenol-D which are predominately useful for military applications. Finally a block **recent and developing** is kind of a bin for the most recent developments like lasers, signal amplifiers, super conductors, water treatment or magnetic refrigeration. Of course this arrangement is of discursive character.

References

1. Hurst C (2010) China's rare earth elements industry: what can the west learn? Fort Leavenworth, Kansas
2. Connelly NG (2005) Nomenclature of inorganic chemistry: IUPAC recommendations 2005. The red book. Royal Society of Chemistry, Cambridge
3. Meschede D (ed) (2006) Gerthsen Physik: die ganze Physik zum 21. Jahrhundert. 23., überarb. Aufl. mit 94 Tabellen, 105 durchgerechneten Beispielen und 1074 Aufgaben mit vollständigen Lösungen und Visualisierungen der Relativitätstheorie auf CD-ROM. Springer, Berlin
4. Mortimer CE, Müller U (2007) Chemie: Das Basiswissen der Chemie; 126 Tabellen, 9th edn. Thieme, Stuttgart
5. Reiners CS (2001) Was ist das Seltene an den Seltenen Erden? Eine chemiedidaktische Reflexion. Chemie in unserer Zeit, 35(2):110–115
6. Allègre C, Manhès G, Lewin É (2001) Chemical composition of the Earth and the volatility control on planetary genetics. Earth Planet Sci Lett 185(1–2):49–69. doi:10.1016/S0012-821X(00)00359-9
7. Cameron A (1973), Abundances of the elements in the solar system. Space Sci Rev 15(1). doi:10.1007/BF00172440
8. Pagel B (1973) Stellar and solar abundances. Space Sci Rev 15(1). doi: 10.1007/BF00172434
9. Rudnick RL, Gao S (2003) Composition of the continental crust. In: Holland HD, Turekian KK (eds)Treatise on geochemistry (pp 1–64). Pergamon [Imprint], San Diego, Science and Technology Books. Retrieved from http://gmg.unizar.es/gmgweb/Asignaturas/Geologia-GradoFisica/apuntes/RudnickGaR2003.pdf. Accessed 11 Nov 2012
10. Binder HH (1999) Lexikon der chemischen Elemente. Das Periodensystem in Fakten, Zahlen und Daten. S. Hirzel, Stuttgart
11. Skinner BJ (1976) A second iron age ahead?: the distribution of chemical elements in the Earth's crust sets natural limits to man's supply of metals that are much more important to the future of society than limits on energy. Am Sci 64(3):258–269
12. Okrusch M, Matthes S (2009) Mineralogie: Eine Einführung in die spezielle Mineralogie, Petrologie und Lagerstättenkunde, 8th edn. Springer, Berlin
13. Anderson DL (2007) New theory of the Earth. University Press, Cambridge
14. Mason B, Moore CB, Hintermaier-Erhard G (1985) Grundzüge der Geochemie. Mit 66 Tabellen. Enke, Stuttgart
15. Willbold M, Elliott T, Moorbath S (2011) The tungsten isotopic composition of the Earth's mantle before the terminal bombardment. Nature 477(7363):195–198. doi:10.1038/nature10399
16. Emsley J (1994) Die Elemente, 2nd edn. de Gruyter, Berlin
17. Roskill (2007) The economics of rare Earths and Yttrium. 13th edn London
18. Hedrick JB (1989) Rare-Earth minerals and metals. In: Bureau of Mines (ed) Minerals yearbook 1989 (pp 825–844)
19. Stosch HG (2002) Geochemie der Seltenen Erden: Skript zur Vorlesung am Mineralogisch-Petrographischen Institut der Universität zu Köln, 1988–1993, Skript mit Ergänzungen von 1998 und Sommer 2000. Retrieved from http://institut-seltene-erden.org/wp-content/uploads/2012/03/uni-koeln-geochemie-der-seltenen-erden.pdf. Accessed 11 Nov 2012. Newer versions are available but not directly retrievable on the internet

20. Röhr C (2011) Chemie der Metalle, Kap. 7.1: Vorlesung. Online Skript. Retrieved from http://ruby.chemie.uni-freiburg.de/Vorlesung/metalle_7_1.html. Accessed 11 Nov 2012
21. USGS (2011b) Rare Earths. Minerals yearbook 2009. Retrieved from http://minerals.usgs.gov/minerals/pubs/commodity/rare_earths/myb1-2009-raree.pdf. Accessed 11 Nov 2012
22. Mortimer CE, Müller U (2001) Chemie: Das Basiswissen der Chemie; 123 Tabellen, 7th edn. Thieme, Stuttgart
23. RÖMPP Online (2011) Thieme chemistry online encyclopedia. Retrieved from http://www.roempp.com/prod/. Accessed 11 Nov 2012
24. von Welsbach C -A (1884) Über die seltenen Erden. Aus dem Universitätslaboratorium des Prof. A. Lieben. Vorgelegt in der Sitzung am 17 July 1884
25. Plohn R (1929) Seltene Erden. Das Lebenswerk Auer von Welsbachs, Die Koralle
26. Greinacher E (1981) History of rare Earth applications, rare Earth market today. In: ACS symposium series pp 3–17
27. Trueb LF (1996) Die chemischen Elemente: Ein Streifzug durch das Periodensystem. Hirzel, Stuttgart
28. Brock WH (2007) Radiant spectroscopy. The Rare Earths crusade. Historical group. Occasional papers no 5. London
29. Pohl WL (2005) Mineralische und Energie-Rohstoffe: Eine Einführung zur Entstehung und nachhaltigen Nutzung von Lagerstätten. Schweizerbart, Stuttgart
30. Markl G, Marks M (2004) Minerale und Gesteine: Eigenschaften, Bildung, Untersuchung. Elsevier/ Spektrum Akademischer, München
31. Schröcke H (1986) Die Entstehung der endogenen Erzlagerstätten. Walter de Gruyter, Berlin
32. Kanazawa Y, Kamitani M (2006) Rare Earth minerals and resources in the World. J Alloy Compd, Elsevier 408–412:1339–1343. doi:10.1016/j.jallcom.2005.04.033
33. Jones AP, Wall F, Williams CT (eds) (1996) Rare earth minerals. Chemistry, origin and ore deposits. Chapman & Hall, London
34. Orris GJ, Grauch RI (2002) Rare Earth element mines, deposits, and occurrences. USGS Open-File Report 02-189
35. Elsner H (2010) Assessment manual: heavy minerals of economic importance. (Ilmenite and Leucoxene, Rutile, Zircon, Monazite and Xenotime, Kyanite, Sillimanite and Andalusite, Staurolite, Garnet, Chromite, Magnetite, Cassiterite, Columbite-Tantalite, Wolframite and Scheelite (2010). Retrieved from http://www.bgr.bund.de/DE/Themen/Min_rohstoffe/Downloads/heavy-minerals-economic-importance.pdf?__blob=publicationFile&v=5. Accessed 11 Nov 2012
36. USGS (2002b) Rare Earth elements—critical resources for high technology. USGS fact sheet 087-02. Retrieved from http://pubs.usgs.gov/fs/2002/fs087-02/. Accessed 11 Nov 2012
37. Bureau of Mines (ed) (1955) Minerals yearbook metals and minerals (except fuels) 1952 (Year 1952, vol 1). Retrieved from http://digicoll.library.wisc.edu/cgi-bin/EcoNatRes/EcoNatRes-idx?id=EcoNatRes.MinYB1952v1. Accessed 11 Nov 2012
38. Kato Y, Fujinaga K, Nakamura K, Takaya Y, Kitamura K, Ohta J (2011) Deep-sea mud in the Pacific Ocean as a potential resource for rare-Earth elements. Nature Geosci 4(8):535–539. doi:10.1038/ngeo1185
39. Gupta CK, Krishnamurthy N (2005) Extractive metallurgy of rare Earths. CRC Press, Boca Raton
40. Jackson WD, Christiansen G (1993) International strategic minerals inventory summary report—rare-Earth oxides. Prepared as a cooperative effort among earth-science and mineral-resource agencies of Australia, Canada, The Federal republic of Germany, The Republic of South Africa, the United Kingdom, and the United States of America. U. S. Geological Survey Circular: 930-N. Retrieved from http://pubs.er.usgs.gov/publication/cir930N. Accessed 11 Nov 2012
41. Berger VI, Singer DA, Orris GJ (2009) Carbonatites of the World, explored deposits of Nb and REE—database and grade and tonnage models. Open-file report 2009-1139. Retrieved from http://pubs.usgs.gov/of/2009/1139/. Accessed 11 Nov 2012

42. USGS (2011a) Mineral commodity summaries. Rare Earths. January 2011. Retrieved from http://minerals.usgs.gov/minerals/pubs/commodity/rare_earths/mcs-2011-raree.pdf. Accessed 11 Nov 2012
43. Steinbach V, Elsner H, Liedtke M (2011) Seltene Erden—Überblick und Perspektiven. Presentation for the german department of foreign affairs. Retrieved from http://zsn.bgr.de/biblioserver/index.php?m=word&kid=2322&gid=1&id=. Accessed 11 Nov 2012
44. Chen Z (2010) Outline on the development and policies of China rare Earth industry. Retrieved from http://www.cs-re.org.cn/en/modules.php?name=news&file=article&sid=35. Accessed 11 Nov 2012
45. Neßler J (2007) Giant Bayan Obo REE-Fe-Nb ore deposit of inner Mongolia, China. Retrieved from http://www.geo.tu-freiberg.de/oberseminar/os06_07/J%F6rg%20Ne%DFler.pdf. Accessed 10 Mar 2012
46. Wu C, Yuan Z, Bai G (1996) Rare Earth deposits in China. In: Jones AP, Wall F, Williams CT (eds) Rare Earth minerals. Chemistry, origin and ore deposits (pp 281–310). Chapman & Hall, London
47. Xie Y, Hou Z, Xu J, Yuan Z, Bai G, Li X (2006) Discovery of Cu-Zn, Cu-Sn intermetallic minerals and its significance for genesis of the Mianning-Dechang REE Metallogenic Belt, Sichuan Province. China. Sci in China Series D 49(6):597–603. doi:10.1007/s11430-006-0597-9
48. Xie Y, Xu J, Chen W, He J, Hou Z, Xu, W (2005) Characteristics of carbonatite fluid in the Maoniuping REE deposit, Mianning, China. Earth and Environmental Science. Mineral deposit research: meeting the global challenge. Session 9, 1097–1100. doi:10.1007/3-540-27946-6_280
49. Gschneidner KA Jr (2011) The rare Earth crisis—the supply/demand situation for 2010–2015. Retrieved from http://www.sigmaaldrich.com/content/dam/sigma-aldrich/articles/material-matters/pdf/the-rare-earth-crisis.pdf. Accessed 11 Nov 2012
50. Yang R, Wang W, Zhang X, Liu L, Wei H, Bao M, Wang J (2008) A new type of rare earth elements deposit in weathering crust of Permian basalt in western Guizhou, NW China. J Rare Earths 26(5):753–759. Retrieved from http://www.sciencedirect.com/science/article/pii/S1002072108601775. Accessed 11 Nov 2012
51. Weixuan F, Shefeng Y, Zhengao L, Xinglin W, Baochen Z (2007) Geochemical characteristics and significance of major elements, Trace elements and REE in mineralized Altered Rpcks of Large-Scale Tsagaan Suvarga Cu-Mo porphyry Deposit in Mongolia. J Rare Earths 25(6):759–769
52. IREL (2012) Indian rare earths limited. Unit profile. Retrieved from http://www.irel.gov.in/scripts/unit.asp. Accessed 11 Nov 2012
53. Paschoa AS (1997)Potential environmental and regulatory implications of naturally occurring radioactive materials (NORM). Appl Radiat Isot (pp 189–196). Retrieved from http://pdn.sciencedirect.com/science?_ob=MiamiImageURL&_cid=271878&_user=616134&_pii=S096980439700239X&_check=y&_origin=article&_zone=toolbar&_ccoverDat=31-Mar-1998&view=c&originContentFamily=serial&wchp=dGLbVlV-zSkWb&md5=1ef9858bd412b0b21f90d2249df5b0ab/1-s2.0-S096980439700239X-main.pdf. Accessed 11 Nov 2012
54. Adunka R (2000) Carl Auer von Welsbach. Das Lebenswerk eines österreichischen Genies. Plus Lucis 1:24–26
55. da Costa Lauria D, Rochedo ERR (2005) The legacy of monazite processing in Brazil. Radiat Prot Dosimetry 114(4):546–550. doi:10.1093/rpd/nci303
56. INB (2012) Indústrias Nucleares do Brasil. Buena—mineral treatment unit. Retrieved from http://www.inb.gov.br/eng/WebForms/Interna2.aspx?secao_id=62. Accessed 11 Nov 2012
57. Rezende MM (2007) Rare Earths 2006. National department of mineral production, Brazil. Retrieved from http://www.dnpm.gov.br/enportal/conteudo.asp?IDSecao=170&IDPagina=1093. Accessed 11 Nov 2012

58. USGS (2007) Rare Earths. Mineral commodity summaries. January 2007. Retrieved from http://minerals.usgs.gov/minerals/pubs/commodity/rare_earths/rareemcs07.pdf. Accessed 11 Nov 2012
59. Chakhmouradian AR, Zaitsev AN (1999) Calcie-amphibole-cliopyroxene rock from the Afrikanda complex, Kola Peninsula, (Russia): mineralogy and a possible link to carbonatites. I. Oxide minerals. Can Mineral 37:177–198
60. Arzamastsev A, Yakovenchuk V, Pakhomovsky Y, Ivanyuk G (2008) The Khibina and Lovozero alkaline massifs. Geology and unique mineralization. 33 IGC excursion guidebook. Retrieved from http://iugs.org/33igc/fileshare/filArkivRoot/coco/FieldGuides/No_47_Excursion_guidebook1.pdf. Accessed 11 Nov 2012
61. Kravchenko SM, Pokrovsky BG (1995) The Tomtor alkaline ultrabasic massif and related REE-Nb deposits, northern Siberia. Econ Geol 90(3):676–689. doi:10.2113/gsecongeo.90.3.676
62. IAEA (2011) Report of the international review mission on the radiation safety aspects of a proposed rare Earths processing facility (the Lynas Project). 29 May–3 June 2011, Malaysia. Retrieved from http://www.iaea.org/newscenter/news/pdf/lynas-report2011.pdf. Accessed 11 Nov 2012
63. Malaysia MTI (2011) The Lynas fact sheet. Official portal of ministry of international trade and industry Malaysia. Retrieved from http://www.miti.gov.my/cms/content.jsp?id=com.tms.cms.article.Article_4008ad57-c0a81573-7a607a60-19ec542e. Accessed 11 Nov 2012
64. Lamb FD, Sargent JD (1958) Minor metals. In: Bureau of Mines (ed). Minerals yearbook metals and minerals (except fuels) 1954 (pp 1383–1395). Retrieved from http://digicoll.library.wisc.edu/cgi-bin/EcoNatRes/EcoNatRes-idx?type=turn&entity=EcoNatRes.MinYB1954v1.p1398&id=EcoNatRes.MinYB1954v1&isize=M. Accessed 11 Nov 2012
65. Carlson EJ, Keiser HD, Sargent JD (1955) Minor metals. In: Bureau of Mines (ed) Minerals yearbook metals and minerals (except fuels) 1952 (Year 1952, Vol I:pp 1172–1185)
66. USGS (2010a) Rare Earths. Minerals yearbook 2007. Advanced release. Retrieved from http://minerals.usgs.gov/minerals/pubs/commodity/rare_earths/myb1-2007-raree.pdf. Accessed 11 Nov 2012
67. USGS (2010b) Rare Earths. Minerals yearbook 2008. Advanced release. Retrieved from http://minerals.usgs.gov/minerals/pubs/commodity/rare_earths/myb1-2008-raree.pdf. Accessed 11 Nov 2012
68. Olson JC, Shawe DR, Pray LC, Sharp WN (1954) Rare-Earth mineral deposits of the mountain pass district, San Bernardino, County California. Science 119(3088):325–326. doi:10.1126/science.119.3088.325
69. Glass JJ, Evans HT Jr, Carron MK, Hildebrand FA (1958) Cerite from mountain pass, San Bernardino County, California. In: The American mineralogist, (vol 43), 460–475. Retrieved from http://www.minsocam.org/ammin/AM43/AM43_460.pdf. Accessed 11 Nov 2012
70. Molycorp (2012) Molycorp's history. Retrieved from http://www.molycorp.com/AboutUs/OurHistory.aspx. Accessed 10 Mar 2012. New link http://www.molycorp.com/about-us/our-history/. Accessed 11 Nov 2012
71. U. S. Environmental Protection Agency (1995) Human health and environmental damages from mining and mineral processing wastes. Technical background document. Retrieved from http://www.epa.gov/osw/nonhaz/industrial/special/mining/minedock/damage/damdam.pdf. Accessed 11 Nov 2012
72. GWMG (2012) Steenkampskraal mine (Rareco), Western Cape, South Africa. Mining. Retrieved from http://www.gwmg.ca/html/projects/mining/index.cfm. Accessed 11 Nov 2012
73. Percy D (2012) Great western minerals group and Ganzhou Qiandong great western minerals group and Ganzhou Qiandong rare Earth group sign rare Earth separation agreement. Saskatoon, Canada: GWMG. Retrieved from http://www.gwmg.ca/html/news/media-releases/index.cfm?ReportID=203424 Accessed 11 Nov 2012
74. Hatch G (2011b) Critical rare Earths: global supply and demand projections and the leading contenders for new sources of supply. Retrieved from http://www.criticalrareearthsreport.com/. Accessed 11 Nov 2012

75. Hatch G (2011a) TMR advanced rare-Earth projects index. Retrieved from http://www.techmetalsresearch.com/metrics-indices/tmr-advanced-rare-earth-projects-index/. Accessed 11 Nov 2012
76. JORC (2004) The JORC code and guidelines. Retrieved from http://www.jorc.org/jorc_code.asp. Accessed 11 Nov 2012
77. USGS (2002a) Introduction to mining. Retrieved from http://media.wiley.com/product_data/excerpt/11/04713485/0471348511.pdf. Accessed 11 Nov 2012
78. NRCAN (2004) Generalized model of mineral resource development. Retrieved from http://mmsd.mms.nrcan.gc.ca/stat–stat/expl–expl/pdf/04_e.pdf. Accessed 11 Nov 2012
79. NRCAN (2006) Mining information kit for aboriginal communities. Retrieved from http://www.nrcan.gc.ca/sites/www.nrcan.gc.ca.minerals-metals/files/pdf/mms-smm/abor-auto/pdf/kit-gui-eng.pdf. Accessed 11 Nov 2012
80. Liedtke M, Elsner H (2009) Seltene Erden. Commodity top news Nr. 31. Retrieved from http://www.bgr.bund.de/DE/Gemeinsames/Produkte/Downloads/Commodity_Top_NNew/Rohstoffwirtschaft/31_erden.pdf?__blob=publicationFile&v=2. Accessed 11 Nov 2012
81. Huang X, Long Z, Li H, Ying W, Zhang G, Xue X (2005) Development of rare Earth hydrometallurgy technology in China. J Rare Earths 23(1):1–4
82. Gschneidner K A, Eyring L (eds) (1978–1996) Handbook on the physics and chemistry of rare Earths. North-Holland, Amsterdam
83. Bünzli J-CG, Pecharsky VK (2011) Handbook on the physics and chemistry of rare Earths. Elsevier Science, Amsterdam
84. Delfrey KN (2008) Rare earths: research and applications. Nova Science Publishers, New York
85. Chegwidden J, Kingsnorth DJ (2010) Rare Earths: facing the uncertainties of supply. 6th international rare Earths conference. Hong Kong, Nov 2010. Retrieved from http://www.slideshare.net/Tehama/rare-earths-facing-the-uncertaintys-of-supply. Accessed 11 Nov 2012
86. Lynas (2010) ASX mid caps conference presentation. March 2010. New York. Retrieved from http://www.asx.com.au/smalltomidcaps/newyork/2010/presentation_lynas.pdf. Accessed 11 Nov 2012
87. Lynas (2011b) Rare Earths. We touch them everyday: Roskill international rare Earths conference presentation. 16. Nov 2011. Retrieved from http://www.lynascorp.com/content/upload/files/Presentations/Final_Roskill_PresentPrese_Nov11.pdf. Accessed 11 Nov 2012
88. USGS and Hedrick JB (2000) Rare Earths. Minerals yearbook 2000. Retrieved from http://minerals.usgs.gov/minerals/pubs/commodity/rare_earths/740400.pdf. Accessed 11 Nov 2012

Chapter 3
The Way to the Chinese Predominance: A Key for Understanding the REE Issue

3.1 The Middle East has its Oil, China has Rare Earths

"The Middle East has its oil, China has rare earth: China's rare earth deposits account for 80 percent of identified global reserves, you can compare the status of these reserves to that of oil in the Middle East: it is of extremely important strategic significance; we must be sure to handle the rare earth issue properly and make the fullest use of our country's advantage in rare earth resources" Ask Metafilter [1]. This quote is attributed to Deng Xiaoping and often cited even though the actual data source is vague. The search for the source leads mainly to Chinese written web pages and news archives. Supposedly Deng Xiaoping talked about the Chinese economic development already during a visit to Inner Mongolia in 1987. Then, during his famous travel through southern China, he supposedly said his famous words in January 1992. In 1999, President Jiang Zemin emphasized during a visit to Baotou the importance of the RE. "Improve the development and applications of rare earth, and change the resource advantage into economic superiority (President Jiang Zemin [2])". This was during the time, when the American Mountain Pass Mine was about to shut down operation. This is remarkable and raises the question, why under such prerequisites the remaining globally important mining in Mountain Pass was being closed; especially when Zemin gives a repeated hint of the importance given by the Chinese to the REE. The quote marks a further step in a strategic plan of China to achieve superiority over the REE. The discussion about this topic probably would be long. One argument which could be said against a master plan is that obviously politicians during visits of significant industry complexes rather strengthen the industry and encourage them to improve their efforts than to remain neutral in their statements. So Zemins' words could also be interpreted as 'no other words would have been expected by a President during the visit of an industrial complex'. If this quote was known to the West, the issue should have given some thought about the handling of possible long term or strategic plans and how a reaction should look like, if such

V. Zepf, *Rare Earth Elements*, Springer Theses,
DOI: 10.1007/978-3-642-35458-8_3,

plans and motivations are made public; meaning, that this sentence of Zemin should have ringed the alarm bell.

In order to better understand China's efforts and actions along the topic of the REE, the objectives of the Five-Year plans give a good insight. Starting with and considering these objectives, the actions of China become (better) comprehensible, and they will show that the President's words fit into a master plan.

3.2 Chinese Plans and Regulations Around REE

Already in 1986 four Chinese scientists suggested to accelerate China's hi-tech development. Deng Xiaoping, who at that time was the Chinese leader, supported the idea which finally was implemented as the National High-Tech R&D Program (863-Program) [3]. One objective was "to achieve 'leap-frog' development in key high-tech fields in which China enjoys relative advantages or should take strategic positions in order to provide high-tech support to fulfill strategic objectives" [3] which influenced Chinese policy in the REE area. The 863-Program is being promoted until today whereas additional programs and plans about the RE industry were set up and added. In 1997 the 973-Program was introduced as a National Basic Research Program [4]. This program again endures until today and sets together with the 863-Program a framework for further specialized plans.

In 2002, Globalsecurity [5] explains that a first attempt to restructure the RE industry failed, mainly due to dissonance at the local level [5]. Hurst [6] gives a more detailed insight into the Chinese effort to achieve knowledge in the field of the rare earths, especially by Prof. Xu Guangxian, who is considered a driving factor and father of the Chinese rare earth chemistry.

3.3 The Five-Year Plan Guidelines, Past and Present

The 11th Five-Year Plan (FYP), 2006–2010, addressed several major objectives which, compared to the 10th FYP, generally pointed on improving already ongoing developments. The objectives were the further growth and quality of the economy as such, a further growth of the national economy itself, the improvement of the income levels of both urban and rural residents, the improvement of public services, the better balancing of urban and rural development, the protection of the environment, enhancement of the market economy system and the progress in the promotion of democracy, legal system, moral and culture. For this, several more detailed cases were raised: first, the expansion of domestic demand, especially consumption to reduce dependence on export. Second, the optimization of industrial structures, the balancing out and upgrading of primary, secondary and tertiary sectors. "Third, we will promote development by relying on resource conservation and environmental protection and focus on the fundamental change of the economic growth mode, transforming economic growth from being driven by large amount of resources consumption to being driven by the improvement of resources utilization efficiency" ([7], p. 1). The further cases show the will to promote more independent innovation, reducing administrative intervention in favor of relying on the market's role and finally the turn towards a more people-centered politics, i.e. improving living standards [7].

A whole lot of plans and intentions then became apparent since 2009 during the period of the 11th FYP. In 2009, China's Ministry of Industry and Information Technology drafted a new plan for the development of the rare earth industry. The core of the plan is the reduction of RE companies to about 20 mergers and thus closing small, illegal and inefficient mines [8]. In September 2010 first plans became public "to promote industrial mergers and acquisitions in a bid to accelerate economic restructuring and increase industrial competitiveness" [9]. In January and February 2011 officials from the MLR announced decisions to set up new standards for the RE sector and to set up 11 state-planned RE mining zones in Ganzhou Prefecture in the ion-adsorption rich area [10]. By this, illegal mining could be stopped and export better controlled. In March 2011, industry leaders of China's largest metals trader called as well for a restructuring of the RE industries, especially in SE China [11]. Then in May 2011, China's State Council issued a national guideline in order to foster a sustainable and healthy development of the Chinese RE industry [12], and a further effort aims on protecting the RE industry from over-exploitation of valuable resources [13]. Bradsher [14] summarizes all this ongoing efforts as a from *Beijing authorities created single government-controlled monopoly* [14]. A set back in the Chinese efforts might be the recent WTO dispute which is discussed later in this document.

The current, 12th FYP (2011–2015), again strongly builds upon the last plan and basically promotes continuing improvements in all previously stated areas. Several reports detail the individual issues, e.g. [15–17]. Four major objective areas are addressed: keyword *change*, including the increase of domestic demand to unleash the huge domestic potential, the improvement and balancing of urban and rural development and the promotion of economic sectors of strategic importance. This shall include the transformation and modernization of seven key industries: clean energy technology, next-generation IT, biotechnology, high-end equipment manufacturing, alternative energy, new materials and clean energy vehicles. Keyword *advance* generally fosters science and technology. Keyword *happiness*, strives to improve living standard and finally keyword *green* makes efforts to reduce massive resource use, reduce environmental degradation by investing in sustainable systems including life-cycle-systems and the development of renewable energies [17]. So the objectives of the 12th FYP underline the need of China for RE products especially in the energy producing sector. Thus the quest for preserving the domestic RE resources gets or remains within an overarching framework.

3.4 Chinese Export Issues Around REE

There are three different instruments active at present: export tariffs, production quota and export quota.

Back in 1985 China began to implement rebates for rare earth exports. In 2004, these rebates were reduced and in 2005 finally cancelled completely. In 2007 export tariffs of 10 % on selected RE products came into effect ([18], p. 8). These were increased until 2011 to 15–25 % for all RE products (oxides, carbonates, chlorides, fluorides, ore and metals) except Praseodymium, Gadolinium, Holmium, Erbium, Thulium, Ytterbium and Lutetium. To this topic, These gives a detailed list of the changes over time [19]. In 2009, there were announcements of members of the MIIT, that a complete export stop for dysprosium and terbium metals is envisaged. That idea however, has not been enforced ([18], p. 8) and can be seen as a media test to determine the uproar and reaction of the rest of the world in order to finally decide for or against such a measure.

In 2008, there was an administrative transfer of the Rare Earth Office from the Chinese Ministry of Land and Resources (MLR) to the Ministry of Industry and Information Technology (MIIT). With that transfer also **production quotas** for REE were set, even though different numbers were issued by the two ministries initially. Eventually in 2010 the numbers were aligned (Fig. 3.1).

So in 2010 there was a production quota of 89,200 t REO (see Table 3.1). Interesting to note in this hindsight is that e.g. the USGS (2011a) and [20] show a production of 120,000 t REO for 2010 which constitutes a strong mismatch. If there is a production quota of 89,200 t REO, how can it be that 120,000 t REO have been produced? Do the 120,000 t REO contain estimated illegal mining data? Are the production quotas not correct? Are they not being enforced? Are the 120,000 t false? In any case a reliable answer to this contradiction could not be found.

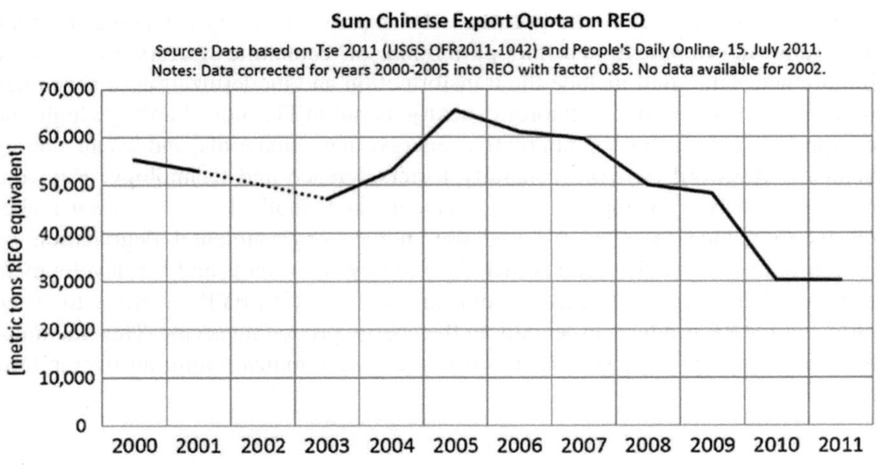

Fig. 3.1 Chinese export quota [19, 27]

Table 3.1 Chinese quota on REE [metric tons REO]

	2000	2001	2002	2003	2004	2005	2006	2007	2008	2009	2010	2011
Production quota[MI]									119,500	110,700	89,200	n/a
Production quota[ML]	55,000	n/a	n/a	n/a	n/a	n/a	86,620	87,020	90,180	87,620	89,200	n/a
Production[e]	73,000	81,000	88,000	92,000	98,000	119,000	133,000	120,000	125,000	129,000	120,000	n/a
Consumption[e]	19,000	20,000	22,000	30,000	34,000	52,000	63,000	73,000	67,700	73,000	77,000	n/a
Export quota[d, 1]	47,000	45,000	n/a	40,000	45,000	48,010	45,000	43,574	34,156	31,310	22,512[2]	14,446[3]
Export quota[f, 1]	n/a	n/a	n/a	n/a	n/a	17,570	16,070	16,069	15,834	16,845	7,746[4]	15,738[5]
Sum export quota	47,000	45,000	n/a	40,000	45,000	65,580	61,070	59,643	49,990	48,155	30,258	30,184
Sum export quota[c]	55,294	52,941	n/a	47,059	52,941	65,580	61,070	59,643	49,990	48,155	30,258	30,184

Notes Data based on Tse [19]: USGS. Open file report 2011-1042. Amended and enhanced

[MI] China Ministry of Industry and Information Technology

[ML] China Ministry of Land and Resources

[c] corrected

[e] estimated

[d] Chinese ventures

[f] Sino-Foreign joint ventures

n/a no data available

blank space no data disseminated

[1] Export quota are given in gross weight rather than REO until 2005. So the data until 2005 have been corrected

[2] Total export quota for 22 domestic producers and traders

[3] Export quota for 2011 is only the first tranche

[4] Total export quota for sino-foreign producers

[5] People's Daily Online [27], 15. July 2011. http://english.people.com.cn/90001/90778/90861/7441312.html

Export quotas have been in place at least since 2000 and have been declined until 2003. In 2004 an increase of the quota was initiated and progressed from about 47,000 t (2003) via 53,000 t (2004) to about 65,000 t (2005) (see Table 3.1). A closer look indicates that in 2004 both export quotas and tax rebates were in place at the same time—if the sources are correct. This remains incomprehensible and answers and explanations are at least not obvious. From 2005 onwards there have been reductions of the export quota together with the allocation of quota to domestic owned and foreign owned RE companies.

The reduction of export quota from 2006 onwards can and has to be seen in accordance with the objectives of the 10th Five-Year plan. "During the 11th Five-Year Plan period, we will implement the basic national policy of resources conservation and environment protection, develop cycling economy vigorously, protect and restore ecosystem and environment, strengthen environmental protection, improve resources management, promote the balanced development of population, resources and environment to realize sustainable development, build a cycling and sustainable national economic system featuring low input, high output, low consumption and emission and a resources-conserving and environment-friendly society" [7]. Chen [18], in an unofficial statement issued by the Chinese Society of Rare Earths, explains some reasons for the Chinese attempts to restrict the export and control the RE business. First he addresses the strong increase in domestic demand, which is completely in accordance with the aims of the five-year plans and the strong economic growth of China. Along follow the intentions of the Chinese rare earth industrial policy to effectively develop and enhance the resources and industry together with the protection of the environment and the promotion of hi-tech application of REE. So the policy covers seven aspects from the licensing of mining, access conditions, export, import, tax and export policy, foreign investment to environmental protection [18]. Keeping in mind that one more aspect of the plans is the strengthening of the domestic industry, i.e. attaining a bigger share of the value-chain, these basic ideas together lead inevitably to the actually implemented actions around the REE. As a summary, Chen comes to the conclusion that "the main purpose of China Rare Earth Industry Policy is to protect the environment, to change the situation of scatter, disorder, and small scale of China rare earth enterprises, so as to elevate the prices of rare earth products to a reasonable range" ([18], p. 9).

With *disorder,* Chen addresses the expected large amount of illegal mining, especially in the southerly provinces, which is estimated to be at 30,000–40,000 t of RE products every year on top of the official production numbers. The word *scatter* points to the large numbers of mainly small scale mining companies. Chen details efforts in this very case [18] and argues that the allocation of export quota showed both a reduction and a concentration of RE industries.

Since about 2009 China is issuing export quotas twice a year for Chinese-owned and foreign-owned RE exporting companies. In 2010 there was a 37 % reduction in export quota compared to 2009. Hatch [21] details the allocation of quota for the second half of 2011 to 23 Chinese-owned and 8 foreign-owned exporting companies [21]. In 2011, the quota nearly remained as for 2010, which is mainly attributed to the assumption that China is reacting to a WTO dispute: in 2009 three member states of the WTO settled disputes against China titled: 'China—Measures related to the Exportation of Raw Materials' (DS 394, DS 395, DS 398). The raw materials covered were various forms of bauxite, coke, fluorspar, magnesium and some others, but no REE. The WTO panel reported about the case on 5. July 2011 [22] and the Chinese Ministry of Commerce announced one day later that China will act according to the relevant laws and WTO rules, however, an appeal is expected (Reuters [28]). On January 30, 2012 the WTO finally issued an Appellate Body Report and basically accepted the pledge and maintained the recommendation that China brings its export duty and quota in accordance with WTO rules [23]. As no steps concerning REE were initiated by China in March 2012 the US, the European Union and Japan issued new disputes related to the Exportation of Rare Earths, Tungsten and Molybdenum, Disputes DS 431, DS 432 and DS 433 [24].

The OECD Trade Policy Working Paper No. 95 as well deals with the impact of various forms of export restrictions on trade in general and global supply. For the case of REE the paper identifies potential caveats for new mining activities as these usually require long-term planning and financing. In the case of REE China has a quasi-monopoly and could change export restrictions as well as tax rebates literally within days, thus rendering potential non-Chinese companies uneconomic [25]. The criticism of the paper is that the arguments which are brought forward as reason for restrictions do not (always) draw expected actions. e.g. if environmental problems are a reason for restrictions, a reduction in production should be the case in order to restore and upgrade facilities; that however, has not (always) been observed. So the OECD claims invalid restriction causes which could lead to trade disputes. Export quota numbers differ in this document from [19]. Numbers for 2000 and 2001 seem to be uncorrected quota numbers, for the data of years thereafter, no explanation could be found. As one data source *Kingsnorth* is given ([25], p. 19, footnote). So in principal the topic and messages seem to be alright; but data differences are present here as well.

Environmental problems finally are not only being addressed by Chinese officials but increasingly getting reported by western media. Hilsum [26] reported from Baotou, Inner Mongolia, where the REE mined from the world's largest rare earth mine in Bayan Obo are processed and refined. The local population complains about poisoned farmland, illegal production and from devastating chemical processing especially in southern China, where acids are reported being flushed directly onto the soil to wash out the valuable REE [26]. By using such primitive methods of mining environmental damages are obvious and remediate actions absolutely necessary.

3.5 Conclusion

China has to improve their domestic RE industry, resolve inefficiencies and solve the intrinsic environmental problems associated with REE processing. As the REE are of strategic, i.e. both extraordinary global and domestic value, China tries to get the RE industry under governmental control. Officially there are three main reasons given for the measures put into effect:

1. The need to reduce the obvious negative environmental impacts and bring the industry up to date,
2. In the same pace stop illegal mining which severely affects prices, efficiency and environment and
3. Supply the strongly increasing domestic demand.

Another side effect is usually mentioned and should not be forgotten as this one has a very strong impact on the minds of China: that China only possess about 50 % of global reserves. That share will probably decrease further as other deposits are discovered throughout the world. So China is afraid of depleting its domestic resources in too short a time, especially the HREE rich deposits in southern China [13]. In hindsight of the expected increase of domestic demand this argument of 'not enough reserves' should not be underestimated. Even though the static range again should not at all raise any concerns. And finally and certainly a further argument can be added for the Chinese politics, that China likes to get more out of the value added chain by selling rather finished products instead of minerals in concentrate form. Irrespective of Chinese reliability in other economic areas, the reasons for the decisive regulations in the case of the REE seem valid and actions absolutely necessary.

References

1. Ask Metafilter (2009) Did Deng Xiaoping say there is oil in the middle east there is rare earth in China? blog about when, where and what did Deng Xiaoping say in this context. Retrieved from http://ask.metafilter.com/132042/Did-Deng-Xiaoping-say-There-is-oil-in-the-Middle-East-there-is-rare-earth-in-China. Accessed 11 Nov 2012
2. RE-Zone (w.y.) Rare earth. An introduction. Retrieved from www.rev.cn/en/int.htm. Accessed 11 Nov 2012
3. MOST (1986) National high-tech R&D program (863 program). Retrieved from http://www.most.gov.cn/eng/programmes1/200610/t20061009_36225.htm. Accessed 11 Nov 2012
4. MOST (1997) National basic research program of China (973 program). Retrieved from http://www.most.gov.cn/eng/programmes1/200610/t20061009_36223.htm. Accessed 11 Nov 2012
5. GlobalSecurity (2010) Rare-earth elements (REE) industry. Retrieved from http://www.globalsecurity.org/military/world/china/rare-earth-industry.htm. Accessed 11 Nov 2012
6. Hurst C (2010) China's rare earth elements industry: what can the west learn?. Fort Leavenworth, KS
7. NDRC (2006) The 11th five-year plan: targets, paths and policy orientation: NDRC minister on 11th five-year plan. Ma Kai, Minister, National Development and Reform Commission. Retrieved from http://www.gov.cn/english/2006-03/23/content_234832.htm. Accessed 11 Nov 2012
8. CIIC (2009) China drafts plan to restructure rare earths industry—China.org.cn. Retrieved from http://www.china.org.cn/business/2009-09/02/content_18450681.htm. Accessed 11 Nov 2012
9. Wang G (2010) China's industries face merger, acquisition push. Retrieved from http://english.gov.cn/2010-09/06/content_1696946.htm. Accessed 11 Nov 2012

10. Tang A (2011) State-planned rare earth mining zones aim for sustainable use, environmental protection: Chinese official. Retrieved from http://english.gov.cn/2011-02/11/content_1801398.htm. Accessed 11 Nov 2012
11. Deng S (2011) Industry leader calls for consolidation of China's rare earth industry. Retrieved from http://english.gov.cn/2011-03/05/content_1817243.htm. Accessed 11 Nov 2012
12. Lu A (2011) China issues guideline to promote healthy development of rare earth industry. Retrieved from http://english.gov.cn/2011-05/19/content_1867140.htm. Accessed 11 Nov 2012
13. Chen Z (2011) Rare earth protection plan. Retrieved from http://www.chinadaily.com.cn/opinion/2011-05/28/content_12596658.htm. Accessed 11 Nov 2012
14. Bradsher K (2011) China consolidates grip on rare earths. Retrieved from http://www.nytimes.com/2011/09/16/business/global/china-consolidates-control-of-rare-earth-industry.html?pagewanted=all. Accessed 11 Nov 2012
15. Casey J, Koleski K (2011) Backgrounder: China's 12th five-year plan. Retrieved from http://www.uscc.gov/researchpapers/2011/12th-FiveYearPlan_062811.pdf. Accessed 11 Nov 2012
16. Foster P (2011) China's five-year plan: key points 12th five-year plan details. The telegraph online. Retrieved from http://www.telegraph.co.uk/finance/china-business/8363275/Chinas-five-year-plan-key-points.html. Accessed 11 Nov 2012
17. Wu H (2011) Vortrag über den 12. Fünfjahresplan Chinas von Herrn Botschafter Wu am 25. Mai 2011, Hotel Albrechtshof Berlin. Retrieved from http://www.china-botschaft.de/det/dszl/baogao/t826701.htm. Accessed 11 Nov 2012
18. Chen Z (2010) Outline on the development and policies of China rare earth industry. Retrieved from http://www.cs-re.org.cn/en/modules.php?name=news&file=article&sid=35. Accessed Nov 11 2012
19. Tse P-K (2011) China's rare earth industry: USGS open file report 2011-1042. Retrieved from http://pubs.usgs.gov/of/2011/1042/of2011-1042.pdf. Accessed 11 Nov 2012
20. Chegwidden J, Kingsnorth DJ (2010) Rare earths: facing the uncertainties of supply 6th international rare earths conference. Hong Kong, November 2010. Retrieved from http://www.slideshare.net/Tehama/rare-earths-facing-the-uncertaintys-of-supply. Accessed 11 Nov 2012
21. Hatch G (2011) Chinese rare-earth export quotas for H2-2011. Retrieved from http://www.techmetalsresearch.com/2011/07/chinese-rare-earth-export-quotas-for-h2-2011/. Accessed 11 Nov 2012
22. WTO (2009): WTO dispute settlement—China—measures related to the exportation of various raw materials: DS 394 (EU), DS 395 (US), DS 398 (Mexico). Retrieved from http://www.wto.org/english/tratop_e/dispu_e/find_dispu_cases_e.htm?year=2009&subject=G158&agreement=none&member1=EEC&member2=USA&complainant1=trtr&complainant2=true&respondent1=true&respondent2=true&thirdparty1=true&thithirdpa2=true#results. Accessed 11 Nov 2012
23. WTO (2012a): WTO | dispute settlement—the disputes—DS394. Retrieved from http://www.wto.org/english/tratop_e/dispu_e/cases_e/ds394_e.htm. Accessed 11 Nov 2012
24. WTO (2012b) WTO | Dispute settlement—the disputes—DS431. Retrieved from http://www.wto.org/english/tratop_e/dispu_e/cases_e/ds431_e.htm. Accessed Nov 11 2012
25. Korinek J, Kim J (2010). Export restrictions on strategic raw materials and their impact on trade: OECD trade policy working papers, vol 95, OECD Publishing, Paris
26. Hilsum L (2009) Chinese pay toxic price for a green world. Retrieved from http://www.timesonline.co.uk/tol/news/world/asia/article6946038.ece. Accessed 11 Nov 2012
27. People's Daily Online (2011) Ministry of commerce: rare earths export quota unchanged. China daily, July 15, 2011. Retrieved from http://english.people.com.cn/90001/90778/90861/7441312.html. Accessed 11 Nov 2012
28. Reuters (2011) China to reform rare earth exports after WTO. Retrieved from http://www.reuters.com/article/2011/07/06/us-china-rareearth-idUSTRE7651RZ20110706. Accessed 11 Nov 2012

Chapter 4
Numbers About Rare Earth Elements in the (Scientific) Literature

4.1 A Literature Review

There are numerous reports available about REE in general, mostly from consultants (e.g. [1–3]). Usually the contents of their reports resemble each other with the exception that some reports contain more details than others, some deal more in depth with producing companies, others stay on top level only. Consultants with emphasis on the financial businesses (e.g. Byron, [4] Crystal International Consultants Ltd., [5]) have reports that start with a general overview of the REE followed by a presentation of present and promising companies listed on the stock markets. These reports again are similar. Furthermore there are several papers available from national agencies like the United States Geological Survey (USGS), the German Federal Institute for Geosciences and Natural Resources (BGR) or the British Geological Survey (BGR) dealing with REE on a more scientific level with emphasis on regular reporting. In contents these reports again are similar and contain usually the same data, presented in different colors and graphs.

Special scientific literature about REE in general is however rare. Instead there exists a huge variety of literature to specific topics within the REE arena, like on geology, geochemistry, extractive metallurgy, magnet production, wind energy, logistics, politics, and so on. They all handle their specific topic very thoroughly but usually lack an overarching view and concept. Scientific reports about life cycles etc. are just beginning to evolve (e.g. [6, 7, 33]).

One interesting fact became apparent during the research of this work which on first sight seemed to be obvious and desirable: the data presented in all the reports showed similar supply and demand data. New reports however brought some doubts and led the author to some calculations about plausibility. As the doubts remained, a thorough literature review for 2009 data was performed which showed that indeed the data has principally one origin: the China Rare Earth Information Center (CREIC) together with unspecified discussions with producers and stakeholders [8]. Except for the unspecified personal discussions this in general is the ideal situation. However, in this case, the source is an office in China disseminating

V. Zepf, *Rare Earth Elements*, Springer Theses,
DOI: 10.1007/978-3-642-35458-8_4,

data which cannot be proved for validity and reliability in a pure scientific sense. Thus seemingly many sources actually have one single basic data set which under scientific rules is doubtful. The results from the literature review are shown in Sect. 4.1. The graph shows the reports by author and year and with simple arrows pointing to the referenced data source. It can easily be seen that most of the reports reference the USGS directly (e.g. [7, 36]), by one intermediate step (e.g. [9, 10]) or both together (e.g. [11, 12]). A look at the data sources given by the USGS leads to BCC Research [13] which references the USGS in turn, [8] and [2]. Kingsnorth is the author of Roskills Book 'The Economics of Rare Earths and Yttrium 2007' [14], and he also is member of the Industrial Minerals Company of Australia Pty Ltd (IMCOA), so that the data sources Kingsnorth, Roskill and IMCOA have to be considered as factually one source. Finally it can be seen, that the last arrows point to the China Rare Earth Information Centre (CREIC), which is factually the only data source for production of REE! It could not be determined, where Asian Metals, Metal Pages and Byron get their data from as these reports and subscriptions could not be obtained (see Fig. 4.1).

In summary this situation is, from a scientific point of view, neither reliable nor valid. Yet, all reports and scenarios are based on these figures.

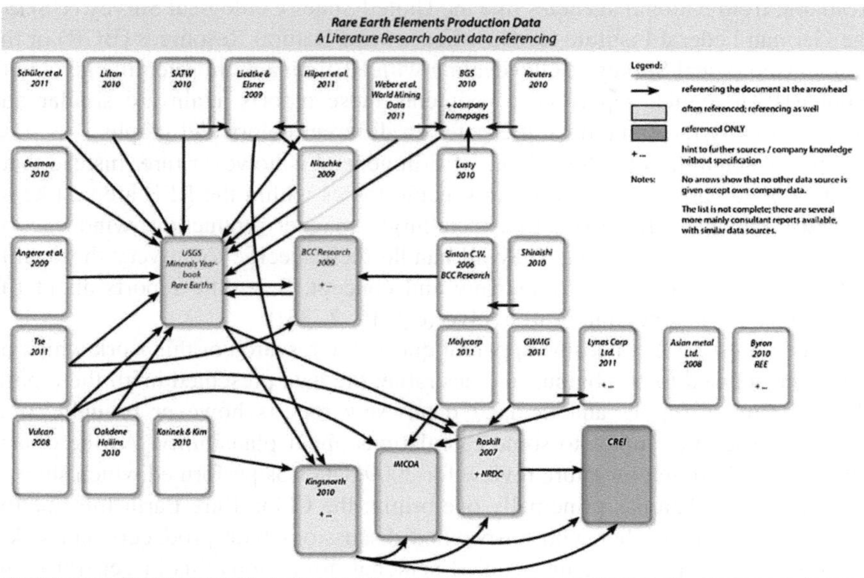

Fig. 4.1 REE production data—literature analysis [29–40]

4.2 Production Data of REE, Cumulative and Individual: An Analysis

First some general problems about data accuracy and reliability are shown before the most probable REE production data are discussed.

4.2.1 Global Annual REE Production: A Comparison of Data Sets Given by the USGS

(Historical) world production data of REE are published by the USGS in basically three ways: as a historical statistics summary, within the Mineral Commodity Summaries (MCS) and as Minerals Yearbooks (MYB). In theory the production data of all three publishing types should be the same. Usually the most recent publications show for the most actual year just estimates which are often revised in succeeding editions. However, within these three types of publications still discrepancies occur. This shall be demonstrated briefly. The common document about Historical Statistics for Mineral and Material Commodities in the United States assembles REE data in the so called 'Data Series 140, Rare Earths' [15] and [16]. Here data starting from 1900 are contained until today in one document, whereas the Minerals yearbook began earliest with production data listing in the 1950s. Before that monazite production data was published (as that was the mined mineral). The historical data set is illustrated in Fig. 4.2.

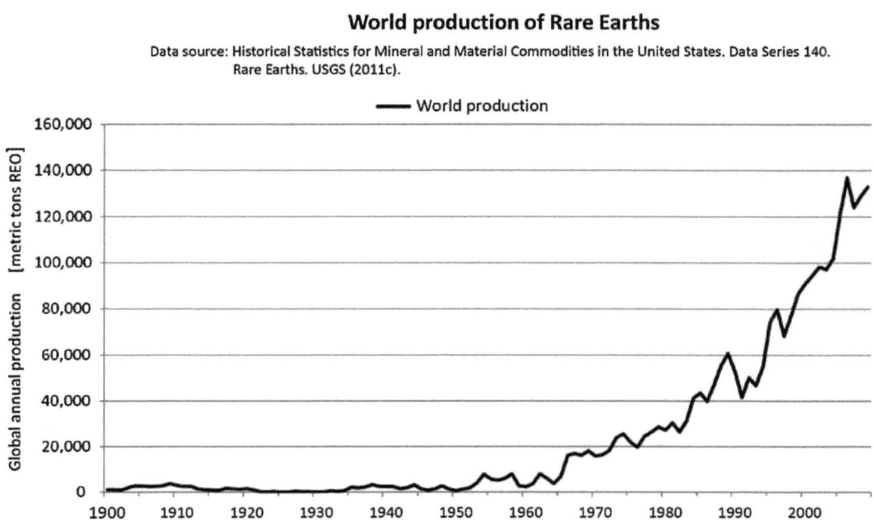

Fig. 4.2 World production of REE, acc. USGS DS140

Figure 4.2 shows the annual global REE production as a measure of REO equivalent quantities from 1900 until 2010. Some decisive information can be extracted. First, the production of REE was relatively low until a considerable production increase occurred around 1953 and another jump around 1966. Both times the increase was about 200 %. The reasons for the 1954 increase are not clearly drawn and may be linked to an increased mining of monazite to attain radioactive thorium and uranium, so that as a by-product REE could be extracted. Furthermore ion-exchange-separation techniques were researched by the AMES laboratory and the demand for intense lighting application and glass manufacturing increased [17]. For 1966 the invention of the color TV using Yttrium and Europium boosted production [18, p. 9]. From 1966 onwards a steadily increasing production began with intermediate peaks in 1989, 1996 and 2006 with subsequent short back drafts thereafter. For the years 1989 and following the decline is explained by [19] with a recession in the US coupled with continued competition. As nearly all areas of REE application showed decreasing production, only the magnets market using neodymium-iron-boron magnets showed some improvement [19, p. 1211]. For the years around 1997 the reason for the decrease in production is explained by declining world economies and stress on thorium-containing compounds which get more and more difficult and expensive to handle [20, p. 61.4] For the 2007 decrease in production no reason is given by the USGS, just that there is a decrease estimated [21].

This data series 140, Rare Earths, lists as data source for the production figures the USGS Minerals yearbooks. So in theory production graphs of both data sets should be the same. This is however not the case. At least there are some discrepancies obvious in the years 1987 until 1994 and 1999 until 2003. This is shown in Fig. 4.3.

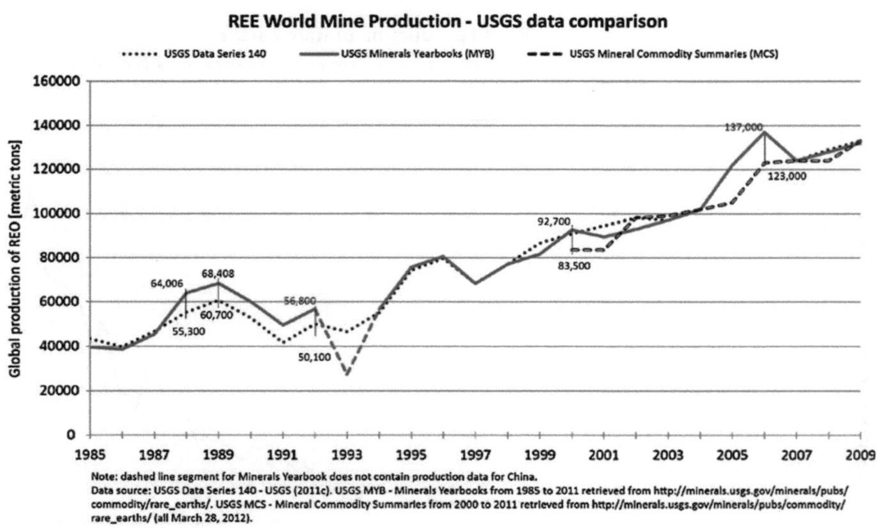

Fig. 4.3 REE world mine production—USGS data comparison

Striking are the discrepancies between the 140s series and the MYB in the years 1988–1992 where differences of around 8,000 t or 15 % exist. A reason for this discrepancy could not be found. Another obvious deviation can be seen between the MYB and MCS data around 2005–2007. For 2006 a difference of 14,000 t REO can be observed which relates to about 11 %. Even though the data in these cases differ, the general trend remains similar, i.e. the two curves actually run nearly in parallel, so that this issue shall no longer be pursued. This could be subject for further research.

It has to be said in this context, that the yearbooks indicate that the data are estimates only. The MCS give a hint as well, usually in the headline of the production tables, that the data are estimated. Nevertheless in succeeding yearbooks it can be traced how the production numbers change and get revised and corrected. Often the record date for the published data is from around mid-year. This seems to be the reason for several data revisions in succeeding years. Still this does not explain the discrepancies noted above. Another questionable point is the data for 1993 where the yearbooks of 1997 state no data available for Chinese production. This is interesting as in the years before and after 1993 the production data for China was available, respectively listed. As well the yearbooks before and thereafter show the respective data. So there remains a general uncertainty in the data. This shall not at all diminish the quality of the USGS data sets but it should illustrate that differing data from even 'one' data source is possible. As a consequence three people could refer to USGS published REO production data for 2001 and all of them have different numbers: 94,500 t for the 140s, 89,500 t in the MYB, and 83,500 t in the MCS. This situation is unsatisfactory and this problem will accompany further considerations done herein. For the remainder of this work, the data from the MCS will be used as they offer the most frequent and regular publication sequence and thus update.

Another discrepancy shall illustrate the data dilemma: Byron [4] shows in their Exhibit 8 of the report the potential global production of important REOs from 2010 to 2015. The data are given for selected REO and as a total sum. For all years, the sum however, is wrong. For 2010 it is obvious that the number for Dy_2O_3 is missing; for the other years it is not obvious, which shares have been omitted or forgotten. The question is why the numbers have been forgotten or omitted and why that has not been explained. It could also be the case, that it was just a mistake in the creation of the table.

In advocacy of the data it can be said, that most of the reports mention that the data are estimates only. However that does not excuse wrong calculations and it does raise the question, of what quality and use these estimated data actually are?

4.2.2 Global Annual Production of REE: A Comparison of Different Authors' Data

Figure 4.4 shows a comparison of eight different production data sets for the years 2000–2008. The columns not drawn lack information. The figure shall illustrate the differences between the authors. The data set for *demand* by [22] has been

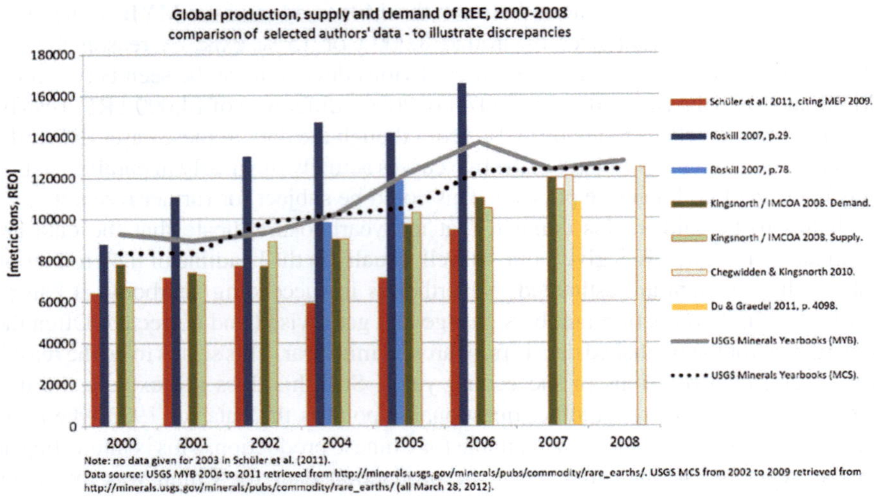

Fig. 4.4 REE global production, supply and demand—a comparison

included in the table to show the balance between supply and demand from that source. This information will be of importance in the following chapters. Most obvious are the differences of the data sets from [7]. Schüler et al., the USGS data set as described above and the Roskill data. Maybe for 2007 it can be said, that all authors show a similar production quantity of about 120,000 t REO.

It is difficult to explain the discrepancies and even a trend is not obvious in the data. The [2] data is the highest and [7] present the lowest numbers. They list several tables with production data of various RE products: RE carbonate, RE chloride, REO, polishing powders, RE metal alloys and single RE metals. As the given source could not be attained, it could not be verified if all REE products have been summed up in the numbers. Another problem is that there is no data given for 2003 without stating a reason. Roskill [2] gives a lot of data in various compilations; however, there are some data discrepancies within the same document: [2] Table 4.2, p. 29, shows the 'World production of RE minerals, 2000–2006' in tons REO with an accuracy of ±15 %. In the same document on p. 78 the output or RE mineral concentrates in tons REO is given which vary from the data in their Table 4.2: 147,000 vs 98,300 t in 2004 and 141,900 vs. 118,700 t in 2005. In 2007 and 2008 the data discrepancies get smaller. However, [8] from Roskill mention that their data include numbers of illegal mining production of about 15,000–20,000 t; these numbers have generally been excluded in and from other data sets. Again there remain doubts about which data have been actually included by Roskill. Have RE minerals and RE mineral concentrates been used synonymously, but if so, why did the differences shrink with time?

As the cumulative data do not show a consistent data basis, and there is no way to determine the correct data set by literature study, the individual share for the REE shall be analyzed.

4.2.3 Annual Production of REE According to Functional Areas

At times, authors give data about the distribution of REE to the major functional or application fields, i.e. catalyst, metal alloys, magnets, phosphors, glass, polishing, ceramics and others. The functional field ceramics is included in *others* here and *battery alloys* is contained in *metallurgy*. As the data for global production showed discrepancies, the same is true for the data sets here. Figure 4.5 shows a selection of data for the years 2005 until 2015 [23, 24].

Figure 4.5 illustrates given and estimated demands by RE applications. The data is arranged group-wise and shows the timeline from 2005 data to the 2015 forecast data. The left column always represents the year 2005 and the most right column represents the estimates for 2015 with a blank column between the years to ease readability. Catalyst data include both automotive and Fluid Catalytic Cracking (FCC) use whereas the metal alloys include battery alloys. The first message—repeatedly—is that there are discrepancies between the different authors. This is obvious in the 2010 data sets (black bordered columns), where the three authors show some considerable differences. The application areas catalyst,

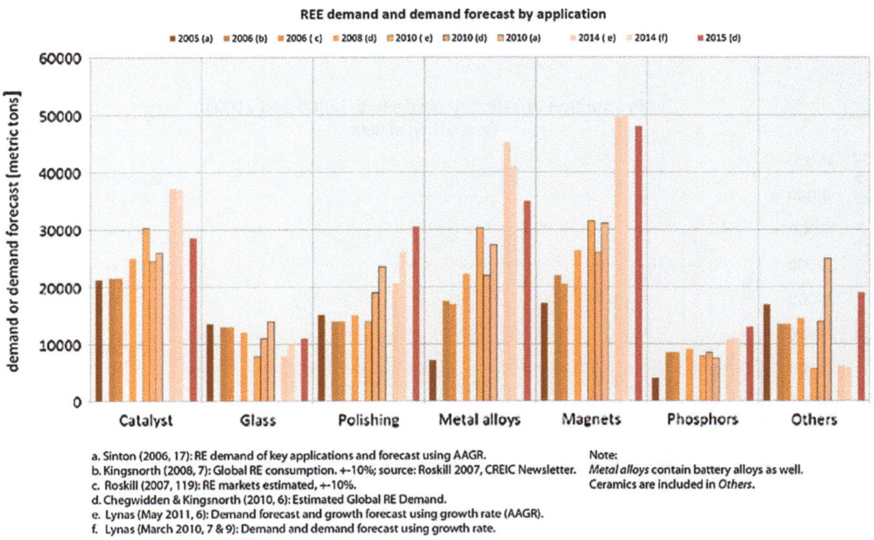

Fig. 4.5 REE demand by applications

polishing, metal alloys and magnets show a general increase in (estimated) demand, whereas phosphors remain approximately steady and glass, ceramics and others show a slight decrease in demand. These estimates were partially derived using AAGRs (see notes in Fig. 4.5) which do not reflect any political, ecological or processing factors. That means that the expected demand is the only driver for supply in this case. This however seems to be changing as China is exerting production and trade restrictions.

Another not so obvious issue is the given data accuracy. Estimates taken in 2007 and 2008 were given as ±10 % accurate; in 2010 the estimates got to ±15 % accuracy. It should be clear, that this 15 % accuracy means that for a given estimated global demand of 185,000 t [8] the spread is between 157,250 and 212,750 t. This overall spread of 55,000 t is in the order of half the annual production in the years 2007 and 2008.

One decisive factor remains hidden in a compilation by application. Each application needs several REE. This is less of a problem for the polishing group where mainly one REE, cerium is used, but decisive for magnets production where mainly neodymium but also praseodymium, dysprosium, terbium and samarium are used. So the data has to be differentiated furthermore.

To get the true picture of demand an individual REE production and use-scheme (including supply and demand) should be considered.

4.2.4 Individual REE Production Share

Individual *supply* data are sometimes given as production data, whereas *demand* data are frequently used in the headlines of respective data. Supply or production

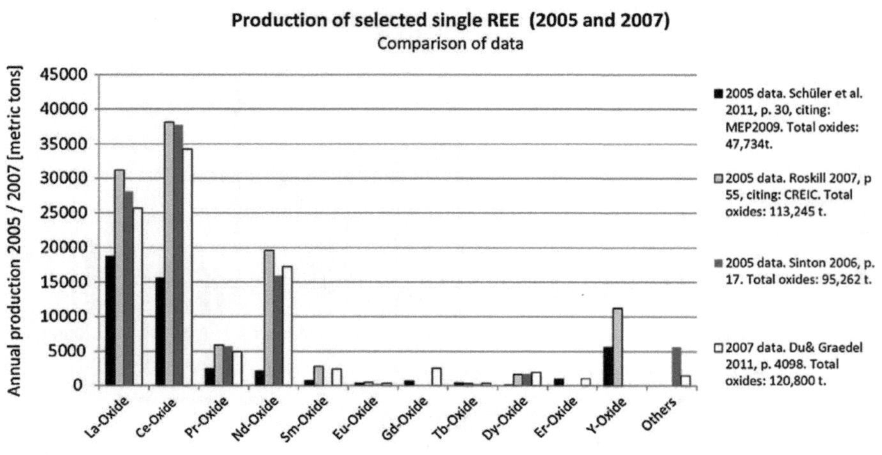

Fig. 4.6 Production of selected single REE ([2, 3, 7, 33])

and demand are of course not the same. As a consequence there is the probability of a data mix up between supply and demand data.

From several data sets it has been tried to set up an overview of individual REE shares in the production. As for the other compilations detailed above, this arrangement shows again considerable discrepancies. Figure 4.6 gives an overview of several REE and their quantitative share in the years 2005 and 2007 given by different authors.

Figure 4.6 shows on the x-axis individual REE from which data was available and on the y-axis the quantities produced in metric tons. The legend lists the data sources, which are indeed partially secondary sources. Next to the three data blocks covering the year 2005 also one data set for 2007 [33] has been chosen. It was decided to include this data as well, due to two reasons: one being that individual REE data was published and that the individual data resembled about the individual data of [2] and [3]. However, Du and Graedel start from an annual production value in 2007 of 120,800 t (Du and Graedel [33], p. 4098) whereas Sinton starts in 2005 with a annual global production of 95,262 t [3] and Roskill states in 2005 about 113,000 t of production [2]. The USGS MYB give similar values for 2005 and 2007 with 122,000 and 124,000 t respectively (see Fig. 4.4 above). The USGS MCS data in comparison show a production increase of 18 % from 105,000 t in 2005 to 124,000 t in 2007. Either way there are unexplained discrepancies and still the three authors Du and Graedel, Roskill and Sinton come up with somehow comparable or similar individual shares. As they start from quite different annual totals there remain some question marks. Nevertheless the three data sets can be regarded as similar in respect of the shares of the individual REE even though the total amount is different; a paradox situation.

In addition to Fig. 4.6 the CSRE also offers production data which resemble the data from Roskill [2, 25]. The data for La are identical whereas the data for Nd differ by about 20 %. This precludes an error of units or conversion, and uncertainty remains.

It can further be seen that the data from [7]—which is supposed to be data from the Chinese official MEP—differs considerably from the other three values. The big difference could not be resolved as the primary source was not attainable. So there remains a big question mark.

Here, a lot more options could be calculated and compared, e.g. the USGS MCS data as a reference base. Instead of elaborating and discussing more options here, the case for Nd in particular will be detailed in more depth and for the other elements mean average values are calculated excluding Schülers' data, as they seem too far off.

The average production quantities for 2005 then are as follows			
La	28,300 t	Ce	36,700 t
Pr	5,500 t	Nd	17,600 t
Dy	1,800 t		

All data refer to REO, rounded and with the explained data reliability

4.2.5 The Quantitative Share of Nd Production

This time and for the remainder of this work, the global production data are taken from the USGS MCS data set. One more data source could act as a proof of at least the Nd data described above. Gutfleisch et al. [26] respectively Luo [27] determine the annual use in 2008 to about 63,000 t NdFeB sintered magnets, based on sales data. On top there are an additional 5,500 t of bonded Nd based magnets so that in total roundabout 70,000 t of magnets were sold. A comparison of this and some other data sets is given in Table 4.1.

The 70,000 t derived from [27] have to be converted to pure Nd and Nd as an oxide which results in 18,500 t of pure Nd (26.7 % of the total quantity) and 21,500 t as an oxide. The [25] gives another data set: "In 2008 about 25,000 tons of neodymium and 31,000 tons of lanthanum were produced. Neither metal is in surplus and supply equaled demand in 2008." [25, p. 1]. Of course this data relates to 2008, but as the global production data for 2007 and 2008 was very similar, it can be assumed that the individual production share was similar as well. From 2005 to 2007/2008 the production data increase about 18 %. With this figure all reported figures can be correlated to get comparable data. It is known, that the different authors have different global production values so that the entire comparison keeps some question marks. Still the spread in production data can be seen. Using the five different authors and the correlated values, the mean average production value for 2005 results in 17,904 t of Nd as oxide. The deviations from the lowest to the highest derived quantities range at around 18 % either side or a total spread about 6,500 t of Nd. For 2008 the differences spread over 8,000 t Nd (−17 – +20 %).

In order to prove this average data, another approach has been used which is to determine the potential individual shares by using the average mineral composition of the biggest Chinese deposits. Of course it is known that an ore body shows quite considerable deviations from the average composition, but this shall be disregarded for the moment. The results of the calculation are also shown in Table 4.2.

For La, Pr, Nd and Dy the differences are below 8 %, for Ce a deviation of 24 % has been observed, i.e. there should have been more Ce produced or reported than actually was. For all other REE a difference of −35 % results which raises some concerns. As the four numbers below 8 % difference could serve as a proof for the validity of the derived average values, the data for Ce and the other REE certainly speak against this idea.

Even though the mineral composition changes sometimes within one ore body, the difference of 24 % seems too big. In other words, assuming the mineral composition is more or less valid there is a considerable amount of Ce missing in demand and a lot of other REE has been produced which is not given by the mineral contents data. A possible answer for the case of Ce could be that Ce actually has been produced in higher quantities than demanded. Consequently Ce has to be available in considerable surplus. Against this theory the price development of Ce in 2011 can be put to argument. Either way the situation is unsatisfactorily. Either the data contain considerable discrepancies or there is a surplus

Table 4.1 Global annual REE and Nd production—a comparison

Global annual production and Nd production—data comparison [metric t REO]

	Data source	2005	2007	2008	2010
Global total production	USGS, MYB	**122,000 t**	**124,000 t**	**128,000 t**	n/a
	USGS, MCS (reference)	**105,000 t** (84.7 % of 2008 value)	**124,000 t**	**124,000 t** (118.1 % of 2005 value)	**133,000 t** (107.2 % of 2008 value)
Global total production	CSRE [25]	21,175 t corr.	25,000 t corr.	**25,000 t**	26,800 t corr.
	Du and Si Graedel [33]	14,620 t corr.	**17,261 t**	17,261 t corr.	18,504 t corr.
	Roskill [2]	**19,600 t**	23,148 t corr.	23,148 t corr.	24,815 t corr.
	Sinton [3]	15,915 t	18,796 t corr.	18,796 t corr.	20,149 t corr.
	Luo [27][a]	18,210 t corr. **[14,000 t]**	21,500 t corr. **[19,800 t]**	**21,500 t [21,500 t]**	21,011 t corr.
	Mean average [excluding bracketed numbers]	17,904 t	21,141 t	21,141 t	22,256 t
	Median	18,210 t	21,500 t	21,500 t	21,011 t

Notes bold figures given numbers

Corr corrected figures, based on given (bold) values and USGS, MCS (reference) percentages

[a] Luo [27] lists global production data of sintered MdFeB magnets for 2005 with 39,610 t; for 2007 with 58,110 t and for 200S with 63,580 t. For 2008, the amount of bonded magnets reached 5,500 t. This value has been added to all 3 years data and corrected for elemental share in magnets (26.7 %) and for oxide contents (85.74 %Nd, 14.26 % O). When the corrected values are compared to Luos given data for 2005, a difference of about 4,000 t becomes obvious. It could not be clarified whether the perceptuali correction or Luos data were in error

Table 4.2 Annual production share based on mineral composition data (Production values for 2005)

	Mineral composition Chinese deposits				Annual production (AP) (t)	Deviation to AV (%)	Average production values (AV) (t)
	Bayan Obo	Sichuan	Others	Grand total			
REO	83 %[a]	3 %[a]	14 %[a]	Σ %[b]	105,000[c]		105,000[c]
La	26	37	15	25	26,030	−8	28,300
Ce	50	47	4	5	45,644	+24	36,700
Pr	5	4	4	5	5,072	−8	5,500
Nd	16	10	15	16	16,4641	−6	17,600
Dy	1	1	7	2	1,932	+7	1,800
Others					9,859	−35	15,100

[a] Chen [28] Share of Mining area on global production
[b] Overall relative percentage of respective REE on global production
[c] Annual production value taken from USGS, MCS for 2005

combined with increasing prices which could be caused by China itself or by speculations. Answers for these discrepancies are outstanding and call for further in depth research.

As no better data (more reliable and valid) could be assembled, these average values for Nd from Table 4.1 will be taken as most probable individual REE data, i.e. 2005: 17,904 t Nd; 2007/2008: 21,141 t Nd; 2010: 22,256 t Nd. These values however will later in this work be questioned again, as they could be grossly wrong or hide a big grey number of applications which are not in the focus today.

Reserves data are not considered herein as the issue has been briefly addressed in Sect. 2.5.

References

1. Lifton J (2010a) The battle over rare earth metals
2. Roskill (2007) The economics of rare earths and yttrium, 13th edn. London
3. Sinton CW (2006) Study of the rare earth resources and markets for the Mt. Weld Complex. For Lynas Corporation Ltd
4. Byron (2010) Electric metals. Green book. April 2010. Retrieved from http://www.google.de/url?sa=t&rct=j&q=byron+electric+metals+green+book&source=web&cd=2&ved=0CDoQFjAB&url=http%3A%2F%2F www.byroncapitalmarkets.com%2Fwp-content%2Fuploads%2F2011%2F02%2F Electric-Metals-Green-Book-April-2010.pdf&ei=zswjT5_DOIer-QaCi73CCA&usg=AFQjCNEcMvN5AUPbxtFdd6VcNxwGexF4sA&cad=rja. Accessed 11 Nov 2012
5. Crystal international consultants Ltd (2011) Webpage. Retrieved from http://crystal-consult.com/de/home/. Accessed 11 Nov 2012
6. SATW (2010) Seltene Metalle: SATW Schrift 41. Retrieved from http://www.satw.ch/publikationen/satwinfo/SelteneMetalle.pdf. Accessed 11 Nov 2012

7. Schüler D, Buchert M, Liu R, Dittrich S, Merz C (2011) Study on rare earths and their recycling: final report for the greens/EFA group in the European parliament. darmstadt. Retrieved from http://www.oeko.de/oekodoc/1112/2011-003-en.pdf. Accessed 11 Nov 2012
8. Kingsnorth DJ, IMCOA (2010) Rare earths: facing new challenges in the new decade
9. Liedtke M, Elsner H (2009) Seltene Erden. Commodity top news Nr. 31. Retrieved from http://www.bgr.bund.de/DE/Gemeinsames/Produkte/Downloads/Commodity_Top_NNew/Rohstoffwirtschaft/31_erden.pdf?__blob=publicationFile&v=2. Accessed 11 Nov 2012
10. Lusty P (2010) Critical metals and rare earth elements. Mines and money 2010. Retrieved from http://www.google.de/url?sa=t&rct=j&q=&esrc=s&source=web&cd=4&ved=0CDsQFjCD&url=http%3A%2F%2F www.bgs.ac.uk%2Fdownloads%2Fstart.cfm Accessed 11 Nov 2012
11. Weber L, Zsak G, Reichl C, Schatz M (2011) World mining data, vol 26. Retrieved from http://www.bmwfj.gv.at/EnergieUndBergbau/PublikationenBergbau/Documents/WMW2011%20mit%20Grafiken.pdf. Accessed 11 Nov 2012
12. Tse P-K (2011) China's rare earth industry: USGS open file report 2011-1042. Retrieved from http://pubs.usgs.gov/of/2011/1042/of2011-1042.pdf. Accessed 11 Nov 2012
13. BCC Research (2009) Rare earths: worldwide markets, applications, technologies: Document No AVM018F. Retrieved from http://www.bccresearch.com/report/rare-earths-markets-applications-avm018f.html?tab=scope&highlightKeyword=AVM018F. Accessed 11 Nov 2012
14. BBY (2010) Conference book rare earths. Retrieved from http://www.bby.com.au/assets/Uploads/Conferences/Rare%20Earth/2010/ConfereCon%20Book%20-%202010%20Rare%20Earth.pdf. Accessed 11 Nov 2012
15. Kelly TD, Matos GR (eds.) (2010) Historical statistics for mineral and material commodities in the United States.: U.S. geological survey data series 140. Retrieved from http://pubs.usgs.gov/ds/2005/140/. Accessed 11 Nov 2012
16. USGS (2011c): Rare earths statistics, 1900-2010: Data series 140. As of Dec 19, 2011. Retrieved from http://minerals.usgs.gov/ds/2005/140/ds140-raree.pdf. Accessed 11 Nov 2012
17. Lamb FD, Sargent JD (1958) Minor metals. In: Bureau of Mines (ed.). Minerals year book metals and minerals (except fuels) 1954 (pp 1383–1395). Retrieved from http://digicoll.library.wisc.edu/cgi-bin/EcoNatRes/EcoNatRes-idx? Accessed 11 Nov 2012
18. Morrison WE (1967) Minerals year book metals, minerals, and fuels 1966: review of the mineral industries. In: Bureau of mines (ed.) Minerals year book metals, minerals, and fuels 1966 (pp 1–76). Retrieved from http://digital.library.wisc.edu/1711.dl/EcoNatRes.MinYB1966v1and2. Accessed 11 Nov 2012
19. Hedrick, J. B. (1991): Rare earths. The Lanthanides, Yttrium, and Scandium. In: Bureau of mines (ed.) Minerals year book 1991 (pp 1211–1237)
20. Hedrick JB (1997) Rare-earth metals in 1997: USGS minerals year book. Retrieved from http://minerals.usgs.gov/minerals/pubs/commodity/rare_earths/740497.pdf. Accessed 11 Nov 2012
21. Hedrick JB (2010) USGS. Rare earths. 2007 minerals year book. Advanced release. Retrieved from http://minerals.usgs.gov/minerals/pubs/commodity/rare_earths/myb1-2007-raree.pdf. Accessed 11 Nov 2012
22. Kingsnorth DJ, IMCOA (2008) Rare earths supply. Alternatives to China: 2008 SME annual meeting
23. Lynas (2010) ASX mid caps conference presentation. New York. Retrieved from http://www.asx.com.au/smalltomidcaps/newyork/2010/presentation_lynas.pdf. Accessed 11 Nov 2012
24. Lynas (2011a) Investor presentation May 2011: rare earths. We touch them everyday. Retrieved from http://www.lynascorp.com/content/upload/files/Presentations/Investor_PresentatioP_May_2011.pdf. Accessed 11 Nov 2012
25. CSRE (2010) The supply of lanthanum and lithium to ford. Retrieved from http://www.cs-re.org.cn/en/modules.php? name = News & file = article & sid = 14. Accessed 11 Nov 2012
26. Gutfleisch O, Willard MA, Brück E, Chen CH, Sankar SG, Ping Liu J (2011) Magnetic materials and devices for the 21st century: stronger, lighter, and more energy efficient. Adv Mater 23:821–842. doi:10.1002/adma.201002180

27. Luo Y (2008) 25 years development of NdFeB magnet industry. Magnews, (Summer and Autumn), 20–24, 22–26. Retrieved from http://www.ukmagsoc.org.uk/uploads/file/Summer%202008%20Magnews.pdf, http://www.ukmagsoc.org.uk/uploads/Autumn2008Magnews.pdf. Both accessed 11 Nov 2012
28. Chen Z (2011) Rare earth protection plan. Retrieved from http://www.chinadaily.com.cn/opinion/2011-05/28/content_12596658.htm. Accessed 11 Nov 2012
29. British Geological Survey (2011) World mineral production 2005–09. Retrieved from http://www.bgs.ac.uk/downloads/start.cfm?id=1987. Accessed 11 Nov 2012
30. Angerer G, Erdmann L, Marscheider-Weidemann F, Scharp M, Lüllmann A, Handke V, Marwede M (2009) Rohstoffe für Zukunftstechnologien: Einfluss des branchenspezifischen Rohstoffbedarfs in rohstoffintensiven Zukunftstechnologien auf die zukünftige Rohstoffnachfrage. Retrieved from http://www.bme.de/fileadmin/bilder/Rohstoffe_fuer_Zukunftstechnologien.pdf. Accessed 11 Nov 2012
31. Asian Metal Ltd. (2009) 2008 annual report on rare earth market. Retrieved from http://www.asianmetal.com/report/en/2008xitu_en.pdf. Accessed 11 Nov 2012
32. Chegwidden J, Kingsnorth DJ (2010) Rare earths: facing the uncertainties of supply: 6th international rare earths conference. Hong Kong, November 2010. Retrieved from http://www.slideshare.net/Tehama/rare-earths-facing-the-uncertaintys-of-supply. Accessed 11 Nov 2012
33. Du X, Graedel TE (2011) Global in-use stocks of the rare earth elements: a first estimate. Environ Sci Technol 45:4096–4101. doi:10.1021/es102836s
34. Hilpert HG, Kröger AE (2011) Seltene Erden—die Vitamine der Industrie. In Mildner S-A (ed.) Konfliktrisiko Rohstoffe? Herausforderungen und Chancen im Umgang mit knappen Ressourcen, pp 159–167. Berlin. Retrieved from http://www.swp-berlin.org/de/publikationen/swp-studien-de/swp-studien-detail/article/knappe_ressourcen_konfliktrisiko_rohstoffe.html. Accessed 11 Nov 2012
35. ISA (2012) (w.y.) Polymetallic nodules. International seabed authority. Retrieved from http://www.isa.org.jm/files/documents/EN/Brochures/ENG7.pdf. Accessed 11 Nov 2012
36. Lifton J (2010b): The supply issue for all metals: the jack lifton report. Special edition. 23 April, 2010
37. Molycorp (2011) Molycorp presentations. Retrieved from http://www.molycorp.com/Resources/Presentations. Accessed 11 Nov 2012
38. Nitschke S (2009) Brief aus Bonn. Forum Metall Trends 2:2–4
39. Reuters (2010) Fight for rare earth. Retrieved from http://graphics.thomsonreuters.com/F/11/RareEarths.pdf. Accessed 11 Nov 2012
40. Vulcan T (2008) Rare earth metals: not so rare, but still valuable. Retrieved from http://www.hardassetsinvestor.com/interviews/1266-rare-earth-metals-not-so-rare-but-still-valuable.html. Accessed 11 Nov 2012

Chapter 5
Rare Earth Elements in the Magnets Application Field

After the data dilemma has been described, that a lot of data and information exists but as well manifold inconsistent data sets are around which are usually not addressed now the focus is put towards the magnet materials to illustrate a new approach to contribute to a partial solution for the problem. First some basics about magnets and magnet technology have to be explained. Probably magnet specialists will call these short explanations trivial but it is expected that even some basic features are not common knowledge to people in other disciplines. Du and Graedel [1] state that typically permanent magnets contain proportions of Nd ($\sim$70 %), Pr ($\sim$25 %), Dy (5 %), Gd (2 %) and Tb (0.2 %). This typical consistency could not be validated, as will be shown by the analysis performed in this work; indeed exactly the many different recipes of the magnets affect the very special features and characteristics. For the supply side it is therefore very important to know the exact quantitative needs in order to start suitable actions. After a short explanation of some terms and basic applications the data of the own research will be detailed.

5.1 REE Based Permanent Magnets

The general use of magnets can be traced back to about 600 B.C. when lodestones were used by the people in the Iron Age. In the Middle Ages iron was detected for its magnetic characteristics and in the advent of industrialization in the 1930s the first AlNiCo magnets were invented. The permanent magnet technology got a decisive advance with the discovery of the strong Samarium-Cobalt (SmCo) magnets in the late 1960s. Two main SmCo type magnets are common: the $SmCo_5$ and the Sm_2Co_{17}. A problem at that time was the low availability and high price of Samarium and, to a lesser extent, of Cobalt. Thereafter, in the early 1980s, the invention of $RE_2TM_{14}B$ magnets (RE—rare earth, TM—transition metal, B—boron) offered a cheaper and better alternative to the SmCo magnets [2, p. 54f]. The RE used in this new type of magnets is generally Nd with the common formula $Nd_2Fe_{14}B$. Numerous patents have been issued in the course until today

V. Zepf, *Rare Earth Elements*, Springer Theses,
DOI: 10.1007/978-3-642-35458-8_5,

with slight deviations in materials composition and manufacturing of the magnets. Soon after these discoveries the RE based permanent magnets were brought into various products, mostly small electric motors but also in hard drives in the computer industry which was growing strongly at that time. Soon the potential of the magnets was obvious so that these strong REE based permanent magnets (REPM) got special attention as they allow better performance on one side and miniaturization which is usually accompanied by smaller product sizes and thus better overall efficiencies.

Some very basic terms and simple explanations are given next as more information is detailed in a variety of literature about (permanent) magnets, e.g. VAC [3], Gutfleisch et al. [4], Fischer [5], Andriessen and Terpstra [6] or Parker [7].

The behavior of a magnetic material inside a magnetic field is a function of the magnetic flux density (B), magnetization (J) and the magnetic field (H) and is characterized in a so called hysteresis-loop diagram. There the relationships can be extracted. The energy density is an energy product of the magnetic flux density (B) and the magnetic field (H). It represents a figure of merit of permanent magnets and has been enhanced continuously until today when about 90 % of the theoretical maximum energy density has been reached. The units for the energy product is MegaGaussOerstedt (MGOe) and starts at around 1 MGOe for steel based magnets to about 56 MGOe for $Nd_2Fe_{14}B$ magnets [4, p. 823]. Furthermore the terms remanence and coercivity are important to note. Remanence represents the remaining flux density, i.e. the magnetism of a material. The coercivity gives the field strength required to demagnetize the magnet completely, i.e. magnetic flux density gets zero ($B = 0$). The higher the coercivity, the stronger is the resistance of the magnet against demagnetization. Besides these characteristics the (operating) temperature can have a decisive influence on the permanent magnet strength or de-magnetization. Next to the susceptibility to electric and magnetic fields, temperature can reversible and irreversibly reduce the remanence. A temperature coefficient indicates the reversible reduction of remanence which will reverse after the temperature decreases. The Curie-temperature indicates above which temperature a magnet will irreversibly loose its magnetic characteristics. Re-magnetization is only possible as long as no changes in the crystal structure of the material have occurred. Several options are possible to enhance the temperature stability, e.g. add new components to the magnetic material, tailor the shape of the magnets or use stabilization processes by applying controlled overtemperature. Nevertheless the demagnetization for NdFeB-magnets starts at temperatures of around 50 °C, depending on magnet type [3]. VAC [3] also suggests 'never-exceed' temperatures of 350 °C for Nd-based magnets and 400 °C for Sm-based PM. For small disc type magnet with 15 × 2 mm a maximum operating temperature of +70 °C is given, for a magnet of the same type with 8 mm thickness +100 °C is possible [8]. As a conclusion not the Curie-Temperature itself is the maximum operating limit but the lower maximum operating temperature. Table 5.1 shows some figures taken from IBS Magnet [8].

The addition of Dy to an NdFeB magnet also enhances the coercivity. Gutfleisch et al. [4] show exemplified the enhanced temperature stability of up to

Table 5.1 Magnet characteristics

	Energy product (B × H max)	Max. operating temperature (°C)	Curie temperature (°C)
Hardferrite (SrFe)	3.4–4.0 MGOe	ca. 200	450
$SmCo_5$	18–20 MGOe	ca. 250	725
Sm_2Co_{17}	20–22 MGOe	ca. 300	750–800
$Nd_2Fe_{14}B$	33–35 MGOe	ca. 120	310

Selection taken from: IBS Magnet [8, p. 4]

175 °C by adding about 4 % Dy as a partial substitution for Nd. Doing this, maximum operating temperatures can be obtained to make these REPM suitable for automotive applications in electric engines for hybrid and electro-vehicles. Sagawa et al. [9] mention that except of La in general all REE could be used as addition to $Nd_2Fe_{14}B$ base magnets but Dy and Tb show by large the best performance and help to enhance the coercivity force. Eventually it depends on the type of application to determine the best performing permanent magnet, i.e. the respective composition, shape and manufacturing. And it should not be forgotten, that hard ferrite magnetic materials do not show this temperature susceptibility, so that they may be alternatives for some uses.

Finally one more problematic characteristic has to be kept in mind. As the REE show a strong electronegativity, they are very reactive and have to be protected against environmental influences to prevent corrosion. Consequently REPM have to be coated for respective uses. Technology for this is either as organic or metallic layer. The metallic layers are usually galvanized using Nickel, Tin or Aluminum coatings and for the organic coatings epoxide resin is applied. Detailed insight is given e.g. by VAC [3]. This topic is essential for potential uses in offshore wind turbines, as the moisture and salty air has to be kept off the magnets.

Most REPM today are manufactured using a sintering process. This allows the production of specially tailored magnets, however common to all REPM magnets is that they are brittle and very susceptible to shock. So special handling care has to be applied.

Overall there are several superior characteristics of REPM versus ferrite magnets, but also some shortfalls. Research is presently widespread in order to overcome the problems and to enhance the even good performance of todays' REPM even more [4, p. 839f; 10, p. 11f].

5.2 REE Based Permanent Magnet Application Overview

A lot of scientific literature is dealing with REPM, mainly as research for improved REPM performance and the uses for REPM. These books obviously explain the topic in much detail which can and will not be repeated here. In order to understand, where REPM actually can be used, more like a list will be assembled to show the huge variety of applications. More detailed insight is given e.g. in Gieras [2] and Fischer [5], but also via homepages of magnet manufacturers like Arnold Magnetic technologies or the German Vacuumschmelze.

REE based permanent magnets are mainly used in electrical engines which in principle are energy converters; either designed as a generator or a motor. In simple terms expressed the generator transforms mechanical energy into electrical, which is e.g. the case in wind turbines. Motors or engines converts electrical to mechanical energy like in an electric vehicle. The construction of either generator or motor is limited to some basic design layouts concerning the stator, usually the casing, and the rotor, a moveable part of the engine. The PM can be attached either to the rotor or the stator with respective advantages and disadvantages so that a variety of combinations of rotor and stator design is possible, both with and without PM, which are detailed in [5]. Now attention is necessary to classify the electrical machines and the PM electric motors without confusing them. Fischer [5] suggests a sorting of electric motors in principle according to the used power type like direct, alternate or rotating current. Furthermore a classification is possible in respect of the mode of operation as asynchronous, synchronous or commutator machines. Gieras [2] offers a classification into 3 different types of electromechanical drives: (1) d.c. commutator drives, (2) brushless motor drives (both d.c. and a.c. synchronous) and (3) stepping motor drives. He then classifies PM electric motors into (1) d.c. brush commutator motors, (2) d.c. brushless motors and (3) a.c. synchronous motors. The sheer number of combinations makes a simple sorting nearly impossible.

It is the question, if a sorted list according to these classifications is helpful. At least it seems useful to have a list of applications which not only name the end use devices but also the production quantities, the material composition of the magnet and the (average) weight of the PM. Such a list with material composition data could reveal the actual big players in the electric drive arena; and, which ones use REE based PM at all. Gieras [2, p. 16] estimates for 2002 a world production of 4.68 billion units of PM motors with a total value of approximately US$40 billion. Luo [11] provides a summary of the development of the NdFeB magnet industry. He mentions that the data gathering is 'not so easy', especially a country-wise collection. Still a list of production is offered starting from 1983, basically the invention of the NdFeB-magnet technology, and reaching until 2008. The table does not contain explicit data sources and cannot be verified. At least some data seem inconsistent to other production data as has been described above. An overview of the share of applications is only available for China so that these data cannot be compared with global application shares [11]. As this suggested overview cannot be provided here just a simple application list shall serve as a beginning, which should be complemented in further research with respective quantitative data.

One of the major problems remains that this list is a probably incomplete list of potential applications, where PM can be used. This does not mean however, that all of these applications actually use PM. And even if they include PM that again does not allow the conclusion, that these are REE based! During the research of this paper also some older power tools, vacuum cleaner and speakers have been briefly investigated; they all had ferrite-type magnets included.

Potential applications of PM electric motors

Note:

- PM does not automatically mean that a REPM is used! Often ferrite or other PM materials are used!
- Information taken, adapted and complemented from [2, p. 19f] and [12].

Industry	• Industrial drives like pumps, fans, blowers, compressors, centrifuges, mills, etc. • Machine tools • Servo drives • Automation processes • Internal transportation systems • (Industrial) robot movement control
Public life	• Heating, ventilating and air conditioning (HVAC) systems • Automatic vending machines • Ticketing and money changing machines • Bar code scanners at supermarkets (actuators for light scanning) • Amusement park equipment
Domestic life/lifestyle	• Kitchen equipment like refrigerators, microwave ovens, dishwashers, mixers, etc. • Bathroom equipment like shavers, hair dryers, electric tooth brushes, etc. • Heat pump type water heaters (ECO Cute) (compressing motors) • Washing machines and dryers • HVAC systems • Vacuum cleaners • Lawn mowers • Toys • Pumps in wells, Jacuzzis, etc. • Vision and sound equipment (mainly speakers in TV, • Stereos; HDD in DVD/HDD recorders) • Cameras (image stabilization, zoom actuators, speakers) • Cell phones (mainly speakers, vibrating motors, sensors to detect open flip) • Portable music players (HDD, speakers) • Security systems like automatic garage doors, gates, etc.
Information and office equipment	• Computers (speakers, HDD with VCM, read/write head, spindle motor, cooling fans) • Printers, plotters, scanners, fax machines, photocopiers (mainly stepping motors)

(continued)

(continued)

Automotive applications	• Ignition coil • Drive motors and generators for HEV, FCV and EV • Electric car air conditioning compressor • Actuator for inter-vehicular distance sensors • Motors for electric pumps • Electric brakes • Starter generator • Electric valve actuators • Seat belt sensors • Car speakers • Car navigation hard disk drives • Motors for electric power steering • Sensors (rotation, angle, position) • Seat actuator motors • Tailgate, door lock and window lift motors • Wiper motors • Fuel pump motor • Sun roof motor • Windshield washer pump
Power tools	• (Cordless) tools like drills, screwdrivers, polishers, sheep shearing handpieces etc.
Medical and healthcare	• Dental handpieces • Electric wheelchairs • MRI (medium strength)
Transportation	• Elevators and escalators • (Electric) Ship propulsion (large PM brushless motors or transverse flux motors above 1 MW) • People movers • E-Bikes, especially in China and increasingly in Europe • Wheel drives • Aerospace systems like HVAC, air compression, etc.
Military and defense	• Weapons (propulsion, actuation of control surfaces,) • Weapon systems like tanks, aircraft, radar systems, submarines, etc. (basically same PM use as for other civil applications)

A quick look into product descriptions of e.g. Kenwood car hifi speakers showed that both RE based and ferrite based PM are used in their products. The same is valid for automobile suppliers which provide small electric motors for seat adjustments, window openers etc. There also RE based and other type PMs are used. The respective share however could not be attained.

It shall be emphasized again, that not only the actual applications of/for REPM is necessary, but also the amount/weight of the magnets used per unit and of course the total number of produced units. Only this can give a solid picture of which applications are the real players and need and deserve respective attention.

References

1. Du X, Graedel TE (2011) Global in-use stocks of the rare earth elements: a first estimate. Environ Sci Technol 45:4096–4101. doi:10.1021/es102836s
2. Gieras JF (2010) Permanent magnet motor technology: design and applications, 3rd edn. CRC Press, Boca Raton
3. VAC (2007) Selten-Erd-Dauermagnete: VACODYM. VACOMAX. Der Fortschritt beginnt beim Werkstoff. Hanau. Retrieved from www.vacuumschmelze.com. Accessed 11 Nov 2012
4. Gutfleisch O, Willard MA, Brück E, Chen CH, Sankar SG, Ping Liu J (2011) Magnetic materials and devices for the 21st century: stronger, lighter, and more energy efficient. Adv Mater 23:821–842. doi:10.1002/adma.201002180
5. Fischer R (2011) Elektrische Maschinen, 15th edn. Hanser, München
6. Andriessen F, Terpstra M (1989) Rare earth metals based permanent magnets: a literature study. Elsevier, London
7. Parker RJ (1990) Advances in permanent magnetism. Wiley, New York
8. IBS Magnet (2008) Dauermagnete. Werkstoffe und Systeme. Product catalogue. Berlin. Retrieved from www.ibsmagnet.de. Accessed 11 Nov 2012
9. Sagawa M, Fujimura S, Yamamoto H, Matsuura Y, Hiraga K (1984) Permanent magnet materials based on the rare earth-iron-boron tetragonal compounds. Magn IEEE Trans 20(5):1584–1589. doi:10.1109/TMAG.1984.1063214
10. Rodewald W, Katter M, Reppel GW (2002) Fortschritte bei pulvermetallurgische hergestellten Neodym-Eisen-Bor Magneten: Hagener Symposium Pulvermetallurgie, Band 18. Retrieved from www.vacuumschmelze.de/index.php?id=87. Accessed 11 Nov 2012
11. Luo Y (2008) 25 years development of NdFeB magnet industry. Magnews, Summer, 20–24, Autumn, 22–26. Retrieved from http://www.ukmagsoc.org.uk/uploads/file/Summer%202008 20%Magnews.pdf, http://www.ukmagsoc.org.uk/uploads/Autumn2008 Magnews.pdf. Both Accessed 11 Nov 2012
12. Shinetsu (2012) Shinetsu rare earth magnets. Magnet applications. Retrieved from www.shinetsu-rare-earth-magnet.jp/e/application/www.shinetsu-rare-earth-magnet.jp/e/ application. Accessed 11 Nov 2012

Chapter 6
Scales and Relations: Analysis of REPM Use with Emphasis on the Years from 2000 to 2010

6.1 Methodology

The determination of actual REPM use followed a twofold approach: first, a primary data analysis was initiated by measuring magnets in HDDs, mobile phones and earphones and second, a literature analysis has been done based on company annual reports, consultant and market research institutes to get annual production numbers. Then the primary data sets for HDDs and mobile phones have been offset with the annual production data to get verifiable quantitative data. For the wind turbine and electro-mobility field all data had to be assembled via literature research, company annual reports, consultant and market research reports. Details of the methods and data used are discussed in the respective chapters.

The primary data analysis has been started to determine the actual weight and composition of both magnets in computer hard drive disks (HDD) and mobile phones. First, old HDDs have been gathered from a collection station in Augsburg which pre-sorts recyclable materials, mainly electronic scrap. In order to get a better sample an email request for old, defect and unused HDDs, mobile phones and headphones has been started within Augsburg University in October 2011. As other communal and official mobile phone collections were planned in near future, the request had to be modest in order not to interfere with these other projects. In return about 200 HDDs, 70 mobile phones and 20 earphones have been collected. Out of these 150 HDDs, 50 mobile phones and all earphones have been disassembled; the magnets have been weighed on a Kern 572 electronic scale with accuracies down to 0.01 g. To avoid any interference of the magnets with the scales a distance cup and a cork cube have been used as a separator.

The measurements have been rounded to the tenth of a gram (1 decimal) and this accuracy has been accepted because most of the magnets still had the expected coatings which sometimes remained intact during disassembly but also partially peeled off. It shall also be mentioned that during disassembly caution has to be exercised as during handling the material can produce sparks, the magnet dust can ignite, the magnets can easily break and produce sharp pieces and finally the

V. Zepf, *Rare Earth Elements*, Springer Theses,
DOI: 10.1007/978-3-642-35458-8_6,

coating is extremely sharp and should only be touched with protective gloves and glasses. The magnets showed the expected brittleness; some magnets crumbled during the attempt to loosen them from the metal plating. One more problem that arose was the magnetism which made it difficult to attach the tools at the right position. The usually strong magnetic forces literally pulled the tools firmly to the magnets so that considerable force had to be used to remove the magnets. After the disassembly and weighing of the magnets they were stored together with their metal casings as some of the magnets are so strong, that they can hardly be pulled apart.

In the next step the magnets were examined for their elemental composition at the Fraunhofer Institute for Silicate Research ISC, Analytisches Dienstleistungszentrum, in Würzburg, Bavaria, Germany. There basically two methods were applied: first a raster electron microscope with EDX analysis (REM-EDX) to get a rather qualitative composition of the magnets and a subsequent RFA (Röntgenfluoreszenzanalyse; XRF—X-Ray Fluorescence Analysis) to get more reliable quantitative information. The general intent was to validate the expected chemical composition of $Nd_2Fe_{14}B$, i.e. a share of about 27 % Nd in the magnets. The magnets were prepared for the analysis by heating them up to 500 °C to get the Nd-based magnets over their Curie-temperatures. After this demagnetization the remaining coatings were mechanically removed. Then the magnet samples were analyzed for all metallic elements with more than 1 wt% by means of an REM-EDX; the analysis of Bor was intentionally omitted as Bor cannot be investigated using this method. Thereafter a selection of samples was analyzed using RFA to achieve more reliable quantitative information. At this stage I would like to thank the Fraunhofer ISC and all associated personnel again for their hospitality and thorough explanations and analysis.

6.2 REPM in Hard Disk Drives

Hard Disk Drives (HDDs) contain several permanent magnets inside the chassis. The most dominant magnets are the permanent magnets of the voice coil motor (VCM). These are usually 2 identical kidney-shaped magnets attached to a metal plate each. This array houses and read/write-head assembly. Some 3.5 in. and most 2.5 in. HDDs only contained one magnet. This magnet assembly houses the wirings of the read/write head. These magnets have been analyzed in this paper. Further magnets are built into the read/write head and the spindle motor. These have not been analyzed in detail as they were too tiny (read/write head) or too difficult to extract (spindle motor magnet).

The typical magnets can be seen on Picture 6.1.

The red arrows are pointing to the magnets of the voice coil assembly (upper left), the magnet in the spindle ring (upper center—crumbled magnet ring) and the tiny magnets in the read/write head.

Picture 6.1 Magnets of a 3.5 inch HDD

6.2.1 Data Issues HDDs

Annual production data was partially available via Annual reports of the major producers, however there quantitative production data is not given frequently so that some information had to be interpolated. For the own measurements of course some inaccuracies are inherent which are explained in the respective chapters. These inaccuracies are deemed not relevant for the overall message though.

6.2.2 Historical Development and Global Shipment Statistics

The HDD industry developed together with the evolution of the consumer PC market. The invention of the strong REE based permanent magnets was one basis for the groundbreaking improvements in the HDD development. Next to the storage densities on the disks itself, the strong magnets allowed a smaller manufacturing of the Voice Coil Assemblies which drive the read/write heads. Initially 5.25 in. size disk drives were used which were followed by 3.5 and 2.5 inch systems which both are major in use technologies today. Smaller sizes are available but the major applications are served with 3.5 in. sizes. Western Digital is said to be the biggest producer with 194 Mio. HDDs shipped in Fiscal Year 2010, whereas the 3.5 in. drives accounted for 144.3 Mio. units (59 %), the 2.5 in. drives represent 80.1 Mio. unit sales (41 %). Other form factors are not mentioned in the reports, except that *there are* others [1]. The remaining 3 big HDD manufacturers in the field probably have similar shares of 3.5 and 2.5 inch systems (detailed data could not be derived). Table 6.1 shows selected highlights in the HDD development starting from the invention of the NdFeB magnets until today, when the companies announced several mergers which will result in three remaining top producers at the end of 2011: Western Digital, Seagate and Toshiba.

The HDD industry differentiates the market usually in Client compute, client non-compute and the enterprise segments. Client compute are the desktop and mobile PCs for homes, businesses and multi-user networks. The non-compute area covers external storage systems, consumer electronics like digital video recorders and game consoles. The enterprise market covers the mission critical storage applications [1, p. 4f]. For all segments the 3.5 and 2.5 inch HDD and increasingly SSD (Solid State Disk) systems build the bulk backbone. Figure 6.1 shows the global shipments of HDDs starting in 1986 until 2010. In 1985 about 4.5 Mio. HDDs were shipped. In 1996 the 100 Mio. barrier has been reached and sales grew slightly exponential until 1999 when nearly 200 Mio. units were shipped. In 2002 the growth increased steadily until in 2007 the mark of 500 Mio. units was

Table 6.1 Milestones in HDD technology

Years	Event	Source
1980s	Invention of Nd-Fe-B permanent magnets	
1986	1st voice coil actuator	D
	About 70 firms in HDD industry	I
1988	1st 2.5″ HDD, 20 MB on 2 disks/platters	D
1991	1st disk drive with magneto restrictive head	D
	1st 1.8″ HDD, 21 MB on 1 disk	D
1996	10.000 rpm HDD	D
	26 firms in HDD industry	I
1997	4 major companies produce 80 % of global production	WD
1999	Seagate produces 250 millionth HDD (cumulative)	S
2003	2.5″ HDD in enterprise application	S
2008	Seagate produces 1 billionth HDD	S
	1.5 TB Disk Drives	S
2010	SSD technology on breakthrough	
2011	4 (3) major HDD producers: Western Digital, Seagate, Toshiba and Samsung (Samsung will become part of Seagate end of 2011)	SPR
Trend	Continued (HDD) storage capacity required/forecasted	TAR
	SSD technology to enter server segments with growth potential (no REE needed)	SAR, WDAR

Sources
D Disktrend [88]
I McKendrick [89]
S Seagate [90]
SPR Seagate [91]
TAR Toshiba [92, p. 13ff]
SAR Seagate [6, p. Sf]
WDAR Western Digital [93]

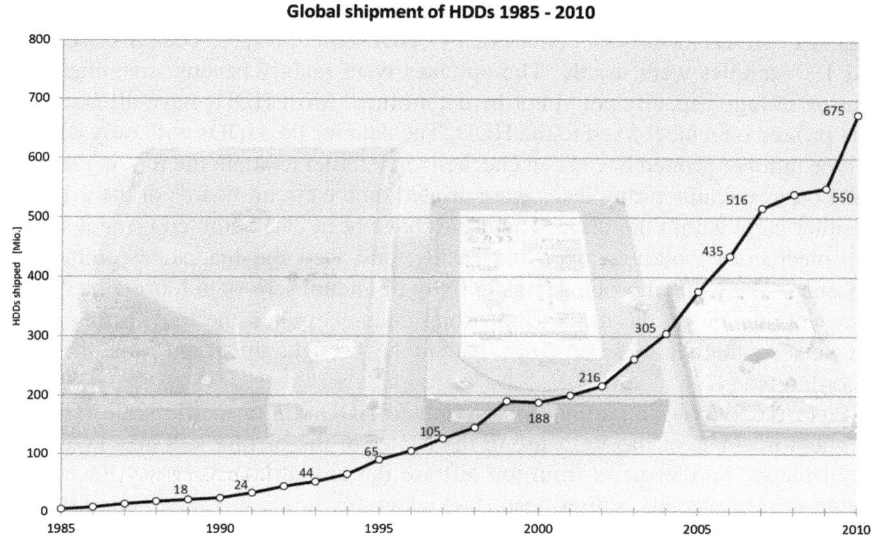

Fig. 6.1 Global shipments of HDDs, 1985–2010. *Data source* 1985–1998: Wong [110]. 1999-2002: Calculated and taken from Western Digital Annual Reports [107–109]. 2003: Trend Focus [105]. 2004–2005: iSuppli [101]. 2006–2008: Western Digital Annual Report [3, 106]. 2009: heise [97]. 2010: iSuppli [100]

surmounted. After two years with modest growth, in 2009 a huge jump happened to reach 674 Mio. HDDs shipped in 2010.

It shall be emphasized that the total shipment data of HDDs has been looked at and not only the shipment data of PCs, as they only account for about a third to a half of the global HDD use.

6.2.3 Measurement of Permanent Magnets Weight in HDDs

At the beginning there were two hypothesizes respectively questions which should be solved. First there was the hypothesis, that with increasing time the advances in hardware would result in smaller permanent magnets inside the HDDs to drive the Voice Coil Assembly. So the quest was to get data on the actual weights and elemental composition of the magnets over time. A literature analysis showed that no satisfying data was available. The only hint was given by Schüler et al. [2, p. 71], where they calculate the amount of Nd used by a measurement of some magnets with an average weight of 22 g [2]. A Request for Information at the HDD producing companies got no satisfying result as well. So the idea was to gather sufficient HDDs to cover about 20 years of development. For this, HDDs from 1990 onwards were collected and disassembled. After the removal of the magnets, the drives were given back to the baskets. All drives have been disassembled either by me or under my supervision and data misuse of the data disks can be precluded. In total, 195 HDDs have been collected, 147 HDDs thereof have been disassembled and 130 samples were usable. The outtakes were mainly because manufacturing year or storage capacity could not be determined. Most HDDs have all necessary data printed on a label fixed to the HDD. The data for the HDDs with only a model or type number printed have been checked via internet to attain the missing data. In some cases manufacturing dates were printed on the circuit boards of the drives or on other parts within the drive. The drives have been disassembled using a special fine mechanics toolkit to open the casing, and then the magnet assembly was loosened. The general problem was initially to find all screws to loosen the casing, then working inside the drives, the strong magnets pulled the tools firmly to the magnets so that quite some force had to be brought up to position the tools accordingly.

Picture 6.2 shows a partially disassembled HDD with a metric scale. The two identical magnets of the VCA are in the center of the picture still attached to the metal plates. Further parts from top left are the circuit board, casing lower side, casing cover, magnets screws head/stack assembly and a 4-disk-storage array.

The next problem was then the separation of the magnets from the securing metal plates. Most of the magnets were only glued with a tiny drop to the metal but together with the magnetic force it proofed very difficult to loosen the magnets. It showed that when cranking too hard, the magnets crumbled and showed the expected porosity of REE based permanent magnets. All the magnets had a coating

Picture 6.2 Disassembled HDD

which could not prevent the magnets from crumbling. Instead the coatings that peeled off were very sharp, so special caution and hand gloves were required.

After disassembly the magnets were weighed on a KERN 572 scales. As has been stated already, the measurement was rounded to 1 decimal because most magnets still contained the coating. So it is obvious that the measured weight is actually too high. It was neglected to calculate and predict the weight of the coatings, as the sizes of the magnets and the different coatings differed very much. This too conservative measurement has to be kept in mind for the final conclusion obviously.

The list of the measurements is attached in Appendix B and Appendix C. A graphical depiction of the measurement results is shown in Fig. 6.2. The figure shows on the primary y-axis a logarithmic depiction of the magnet weight in grams, on the x-axis the production year and on the secondary y-axis (right side) the ratio of weight per storage capacity given in g/GB. Here a clear trend was obvious, that the weight to storage capacity ratio shrinks with time (black dotted line). The magnets of the 3.5 in. HDDs showed weights ranging from 2.5 g for Western Digital WD200 drives (e.g. label 15) to 76.6 g for a 2001 Seagate drive (label C61). A general trend for 3.5 in. drive magnet weight could be demonstrated with a slight decrease in magnet weight over time (black solid line). The average weight of the entire sample is 16.3 g; covering all the years, disregarding any specialties, like SCSI or storage capacities.

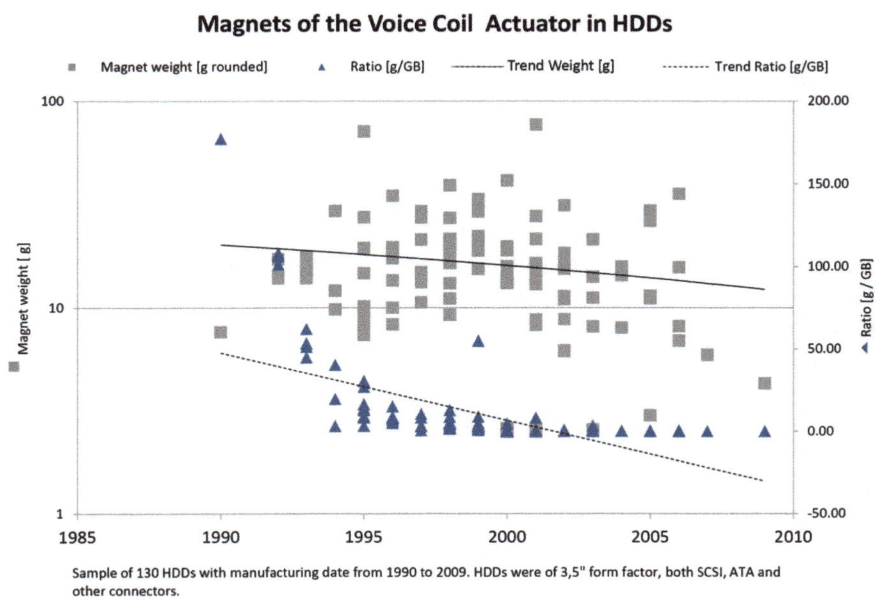

Fig. 6.2 Magnet measurement data of the VCA in HDDs

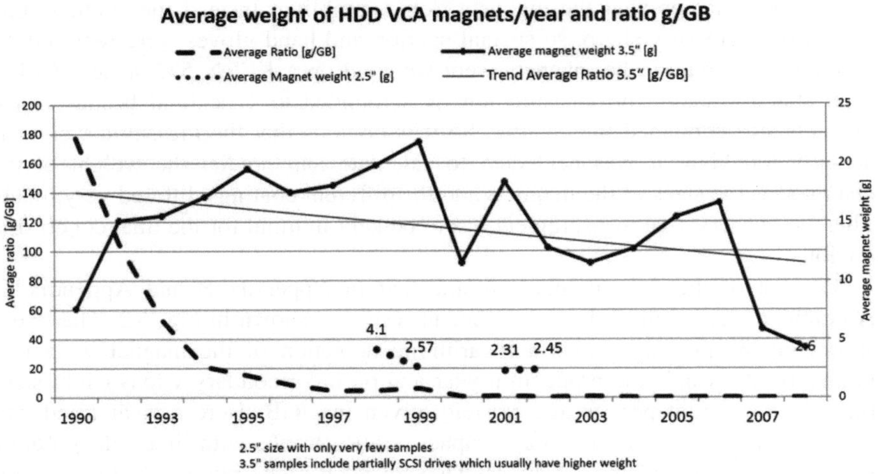

Fig. 6.3 Average weight of HDD VCA magnets/year and ratio g/GB

Figure 6.3 details the average weights and weight/capacity ratio determined by years. The primary y-axis shows the ratio of g/GB which represents an obvious (dashed) decreasing line. The (solid) weight/year line shows an erratic development which maybe could be smoothed if more samples would have been available. Nevertheless the trend which was indicated on Fig. 6.2 can clearly be seen here (black line). Some sparse entries are depicted for 2.5 in. drives with the dotted line. Here no trend could be determined as not enough samples were available. With the exception of one 2.5 in. drive dated 1998, all other drives showed magnet weights of about 2.5 g. For this an estimation of 2.5 g per 2.5 in. HDD could serve as a reasonable value for the intent of this analysis.

The chemical analysis was performed with 10 samples and the results are shown in Fig. 6.4. The idea was to verify the expected weight of Nd according to the formula $Nd_2Fe_{14}B$, i.e. Nd should have a share of 27 %, Fe should have 72 % and B the remaining 1 %. The relative share could be verified, however with some changes in the details. Until about the year 2000 the magnets only contained Nd, from then on Pr was used to substitute about 5 % of the Nd. It is not obvious whether actually the individual elements Nd and Pr were manufactured or whether already the Didymium was used. The samples that were analyzed by RFA showed an average content of 25 % Nd, 5.3 % Pr and 66 % Fe. Interesting to note is that the REM-EDX measurements showed about 5 % more quantitative share for Nd and Pr whereas Fe numbers were about 10 % lower than the RFA values. Here the RFA data are more reliable in this quantitative analysis method than the REM-EDX measurements. Consequently the RFA data have been used for further considerations.

In Figs. 6.5 and 6.6 finally the data of the global annual shipments and the determined average required quantities for the respective years have been

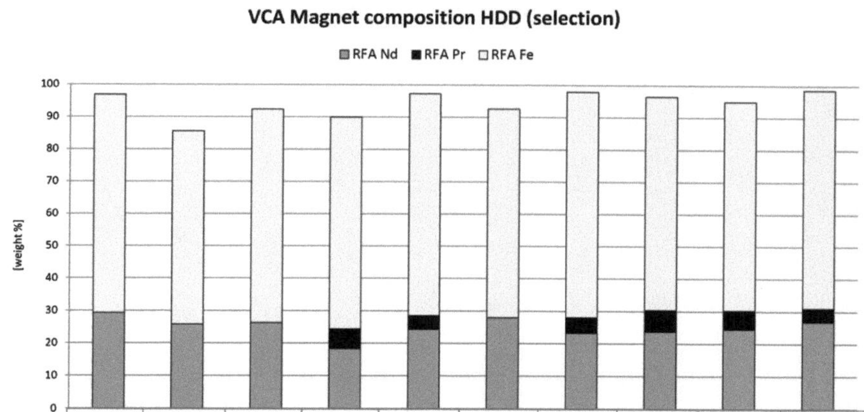

[wt %]	1996 WD Caviar 32100	1998 IBM DTCA 24090	1999 IBM DNES 309170	1999 Toshiba MK8113MAT	2000 Seagate ST320423A	2000 WD WD200	2004 Samsung SP0802N	2007 WD 883D	2009 Seagate ST3500418AS	2009 Seagate Momentus 5400
RFA Fe	67.6	59.8	66.1	65.6	68.7	64.6	69.9	66.1	64.7	67.6
RFA Pr				6.1	4.3		4.8	6.7	5.8	4.2
RFA Nd	29.3	25.8	26.3	18.3	24.3	28.0	23.3	23.7	24.5	26.8

Fig. 6.4 HDD VCA magnet composition

combined. The weight data is purely based on the measured data in Fig. 6.5. The global sales data have been multiplied by the average weights of the respective years to get the quantities required of both Nd and Pr to produce the magnets. Until the year 2005 no 2.5 in. data have been included, from there on the proportions given by Western Digital [3][1] have been assumed for the other manufacturers as well. For the years 2008 and 2010 no selective data for the share of 2.5 in. HDDs was available so that the 2007 value of 28 % has been used instead. It is likely that the actual percentage of the 2.5 in. drives actually was higher so that the calculations done here are maybe too conservative; i.e. in reality less magnet weight has been used. Further research could complement the data. For the case of the 3.5 in. drives not enough samples were available for the years 2007 and thereafter. This is obvious as a HDD should be in use longer than a few years. The few samples could actually be representative but the big jump to a lesser magnet weight is unlikely. Therefore a second graph has been compiled using a mean average magnets weight taken from the years 2000–2006 which equates to 14 g magnet weight/HDD as shown in Fig. 6.6.

Figures 6.5 and 6.6 show that the weight of the pure Nd needed to produce the shipped HDDs started with 54 t in 1990, reached about 524 t in 1995 and had a peak in 1999 when about 1,200 t were needed. In 2000 the shipments declined and a partial substitution of Nd by 5 % of Pr was noted. As a consequence the demand

[1] Following percentages have been reported by Western Digital and used analogue for the other manufacturers: 2005-2 %, 2006-7 %, 2007-13 %, 2008-28 %, 2009–2010-28 % were assumed as no reported data was available. The average weight: 2.5 g per 2.5″ HDD.

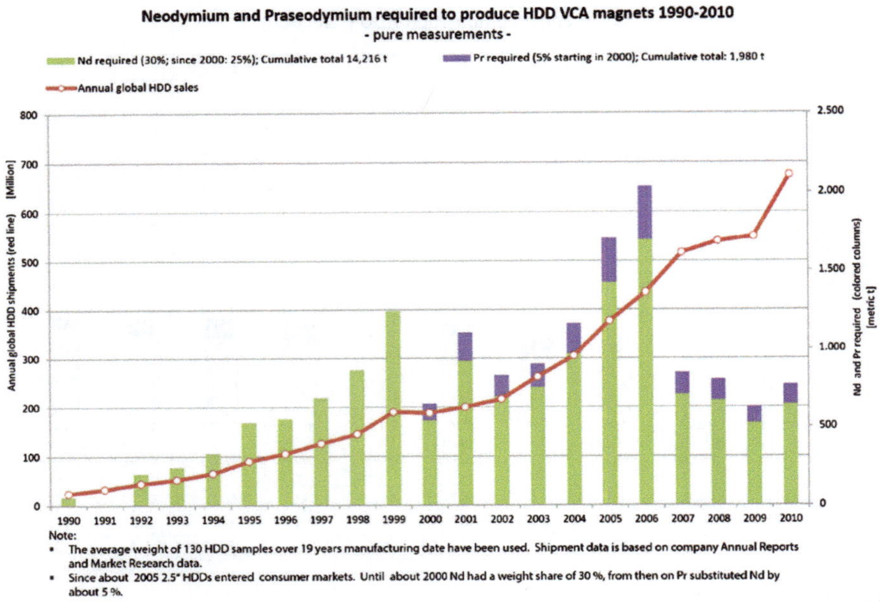

Fig. 6.5 Nd and Pr required to produce HDD VCA magnets 1990–2010 (pure measurement)

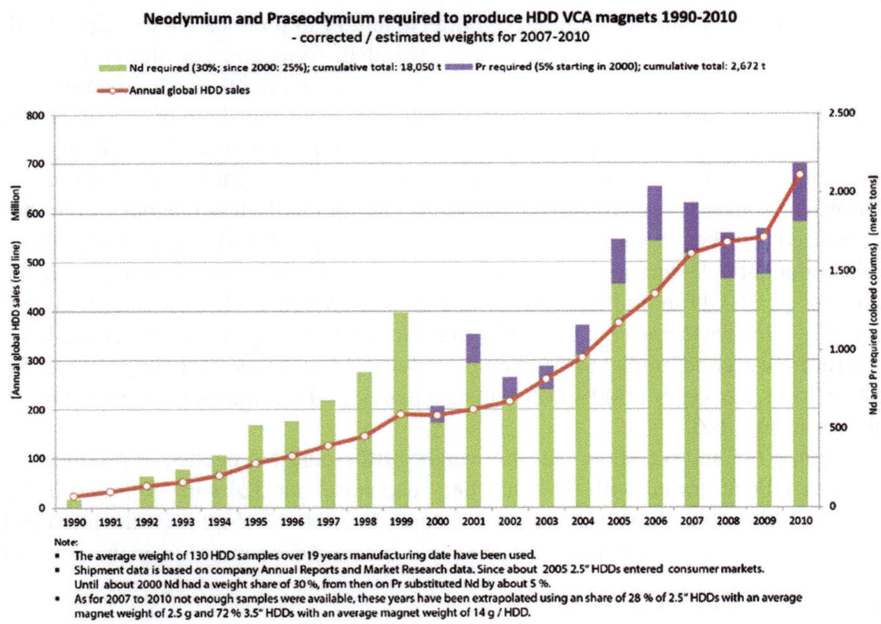

Fig. 6.6 Nd and Pr required to produce HDD VCA magnets 1990–2010 (corrected)

Table 6.2 Nd and Pr required to produce HDD VCA magnets 1991–2010

	Nd [metric t]	Pr [metric t]
1991	n/a	n/a
1992	199	0
1993	241	0
1994	332	0
1995	524	0
1996	551	0
1997	684	0
1998	861	0
1999	1,243	0
2000	540	108
2001	920	184
2002	691	138
2003	752	150
2004	968	194
2005	1,423	285
2006	1,698	340
2007	1,614	323
2008	1,455	291
2009	1,481	296
2010	1,818	364

Note
Nd required based on 30 % wt of Nd in VCA magnets untill 1999, then 25 % wt
Pr required based on 5 % wt since 2000

for Nd in 2000 in this arena diminished to 540 t. After some variable years a new peak was reached in 2005 driven by strong sales which required about 1,400 t Nd and 284 t Pr. In 2005 the 2.5 in. HDDs entered the mass market with an initial modest 2 % share of the global shipments. Eventually in 2006 the highest peak was reached requiring about 1,679 t Nd and 336 t Pr. In 2007 the global sales data continued the trend before until 2009 a shallow increase was observed. In 2010 a jump in sales occurred from about 550 Mio. HDD shipments in 2009 to nearly 675 Mio. in 2010. The share of the 2.5 in. HDDs grew to 28 % in 2008 so that the amount of magnet required did not grow according to the trend of the past years. These data are assembled again in Table 6.2.

Coming back to the earlier determined annual production in 2007 of 21,141 t Nd the HDD magnet use results in a share of 3.3 % (704 t; measurement) respectively 7.6 % (1,614 t average estimation). For 2010 with 22,256 t Nd the 1,818 t Nd corresponds to about 8.2 %. As the weight of the magnets was measured with coatings the numbers are probably slightly high. Nevertheless, further in depth analysis would be required to get more reliable and accurate data. One problem thereby is that the required samples of the few recent years are not readily available—without enough funds to buy and disassemble new HDDs.

6.2.3.1 Spindle and Read/Write Head Magnets

In this whole analysis the spindle and read/write head magnets have been disregarded in the calculations. The magnets of the read/write head are considered too small and not relevant for a quantitative message. Of the magnets in the spindle drive some have been scratched from the housing and measured to give an idea of the weight. A 3.5 in. Western Digital WD Caviar 307AA showed a ring magnet in the spindle with a total weight of 2.7 g; a 3.5 in. Samsung SP 2014 N weighed 2 g. The small 2.5 in. Fujitsu MHT 2080AH had a magnet with 1 g of magnet weight. The REM EDX analysis of the WD Caviar 307AA spindle magnet ring showed a quantitative content of 30 % Nd, 65 % Fe and 5 % Co. Factually for each HDD about 0.6 g more Nd should be added for the total magnet quantity. As a very rough estimate this would add in the order of around 400 t to the annual demand for Nd in HDDs. But this calculation is based on one sample of a 3.5 in. HDD only.

Furthermore some CD- and DVD drives have been disassembled. In average there were the following magnets (weight) included: 2 cubic PM next to the laser (1 g each), 1 ring (washer type) magnet to press the CD/DVD to the spindle (1.1 g), 1 spindle ring magnet (2.7 g), 1 rubber magnet in the traction motor (4.4 g ferrite?) and 2 magnets for positioning the laser array (2.2 g each, ferrite?). One drive, a Lite-On CD-ROM, Model LTN-483L, manufactured in 2001, has been analyzed as a snapshot with REM EDX. The cubic PM magnet next to the laser showed 28 % Nd, 10 % Pr, 61 % Fe and 2 % Co (rounded); the washer type ring magnet showed 42 % Nd and 58 % Fe; the spindle ring showed similar values as the HDD spindle ring: 30 % Nd, 65 % Fe and 5 % Co. Again, these values only represent a more qualitative view and further analysis would be necessary; especially the contents of the rubber magnets and magnets in the positioning motor are of interest as they are the heaviest ones. Consequently for each PC shipped it can be assumed that one drive is built into it. In 2009 Gartner said, that about 308 Mio. PC were shipped [4]. Assuming that each PC contains a CD/DVD drive, there is the need for about 308 t Nd per year alone for these gadgets. On top of these lots of DVD players, stereos, car navigation systems and playstations rely on the same drive technology so that a considerable amount of drives is around.

6.2.4 *Recycling Potential for HDDs*

There certainly is a theoretical recycling potential for the magnets of HDDs; however accompanied by a variety of problems. First of all there is the potential to recycle production scrap during the manufacturing of the magnets themselves. In the Post-consumer-phase, i.e. the recycling of products that reached their end-of-life the challenge is, next to the collection of enough HDDs for recycling, the separation of the magnets from the HDD (casings). A manual separation seems not economical so that specialized automatic techniques are desirable to disassemble the magnets from the HDDs automatically. Hitachi [5] announced that they developed a technology for REE magnet recycling stemming from either HDDs or air conditioners. New cutting machinery helps to extract the magnets which then are dry processed to separate the REE [5]. In how far this technology is in practical use is not known. Next there is the question of how much magnet material would be available for recycling. Using the data from Figs. 6.5 and 6.6 above the global potential results in a cumulative total from 1995 until 2009 of approximately 14,000 t (measured)/ 18,000 t (average estimation) Nd, worldwide. It is doubtful, whether the drives before the year 2000 are still available for recycling so that around 4,000 t less would be available. This value could be compensated by the not considered magnets from the spindle drives.

So a potential quantity of 10,000–14,000 t of Nd globally seems realistic. Western Digital [1] and Seagate [6] estimate in their geographical revenue data that the distribution of HDDs is about 25 % in the North Americas and Europe and about 50 % in Asia–Pacific region. Even though in earlier years this share probably was more in favor of North America and Europe, the amount of potentially recyclable HDDs is regionally limited.

Next to these theoretical figures it has to be clear that not all of these magnets would be available instantaneously but with a lag of several years as the HDDs should be in operation for up to 10 years. Of course new HDDs will be produced but the share of HDDs could be reduced as new storage technologies like SSDs and higher capacity HDDs get on the market.

6.2.5 Outlook HDD

For all three segments client compute, client non-compute and enterprise, a growth in demand is forecasted due to the expected need for more storage capacity in general, i.e. increase in demand in all segments and the growing use of cloud computing systems. That means that a lot of storage capacity might be outsourced from private and corporate systems to the cloud in future. It is thinkable that a rebound effect indeed will lead to an increased storage demand. The fact that storage gets cheaper; easier to use and to access (via the cloud); that applications, picture and video resolutions get better; that the accompanying data streams are getting faster, and that the general population growth asks for more computers, all this could lead to a growing demand both for classic hard disk storage and new solid state storage. The probable quantities however cannot be estimated and even among market analysts disagreement exists about the future demand.

Annual requirement for HDD|VCA magnets		
2005	1,423 t Nd	285 t Pr
2010	1,818 t Nd	364 t Pr
2015	No reliable predictions	
2020	Available	

6.2.6 Further Research

The data derived could be verified and amended by more recent drives as well as the incorporation of more spindle drive magnets. An enhancement would also be interesting as to include CD and DVD drives, both in computing and audio systems. Of course the global production data, the share of 3.5 and 2.5 in. systems and the regional distribution would be of interest for a more accurate picture. Finally the expected growth would be of interest to determine the future demand.

6.3 REPM Use in Mobile Phones and Earphones

The methodology for analyzing mobile phone REPM use was applied analogous to the HDD analysis. First a literature review assembled global sales data starting in 1997 until 2010. The second step was the collection of used mobile phones followed by the disassembly of the speaker and vibration motor magnets. The weight was determined as for the HDD magnets, and some samples have been analyzed at the Fraunhofer ISC using REM-EDX and RFA. Together with the sales data, the magnets weight and composition data the annual usage of Nd and Pr has been determined.

6.3.1 Data Issues Mobile Phones

Here again the global sales data had to be used from analyst publications and could not be verified in total. In general the sample is probably too small as hundreds of different models have been produced. As for the HDD no most recent models of Smart Phones were available for disassembly.

Picture 6.3 shows the typical magnets built into a mobile phone: left top the vibration motor, below one speaker magnet (actually the silver ring under the golden shimmering ring and below the tablet shaped magnet of another speaker.

6.3.2 PM Use in Mobile Phone Systems

In a mobile phone there are principally 3 applications[2] for REPM: one being in the speaker of the phone, one within the vibrating motor and 2 Magnets built into the headset which is usually shipped with a mobile phone (see Picture 6.3). The

[2] The need for REE like Ce for the glass of the displays or Eu or Tb for the LEDs is not considered herein.

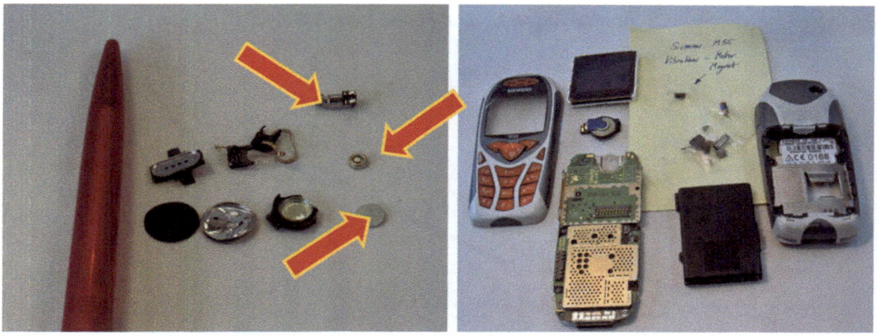

Picture 6.3 Disassembled mobile phone

speaker contains a small magnet either as a ring or in tablet forms with about 3 mm diameter. The magnet is attached to a washer for protection and fixed inside the speaker array. Most mobile phones use one speaker; sometimes two speakers are used, preferably in music-oriented (Walkman) phones like the Sony-Ericsson W580i or generally in slider and flap systems. The weight of these magnets varies from 0.16 g (Siemens M35i) to 0.72 g (Alcatel BE1). In average the weight is around 0.3 g. The vibration motor occurs in two basic designs: a tablet form and in a cylindrical shape. The weight of the entire motor is in the range of 1.5 g, the magnets and magnetic parts were measured from 0.04 g (Siemens ME45) to 0.31 g (Samsung SGH X800). The problem of the measurement was the disassembly of the magnets from the tiny motors which only succeeded in a few cases. The headsets which are shipped together with most phones contain two phone speaker magnets (left and right speaker) either in ring or disc shape. The weight of these two magnets together was 0.56 g in average, ranging from 0.18 g (Elecon, in ear) to 0.7 g (AKG K315, Hi End, in ear) total weight for both magnets. As a summary the average weight of these magnets per mobile phone including one headset is 1.1 g. From this 1.1 g in average about 26 % are Nd, i.e. 0.28 g per mobile phone and about 5 % are Pr, i.e. 0.06 g per phone. Indeed the average weight is most probably a little lower as not all earphones had a REPM but a ferrite magnet built in.

6.3.3 Historical Development of Global Annual Mobile Phone Sales

Figure 6.7 shows the development of the annual global sales from 1997 to 2010. The data was extracted out of several press releases from Gartner Research, which are available free on the internet. In 1997, 108 Mio. mobile phones were sold and a strong growing trend continued until 2000 when sales stagnated, before in 2003 a steady growth continued until about 2008. In 2009 a big step from 1,200 Mio.

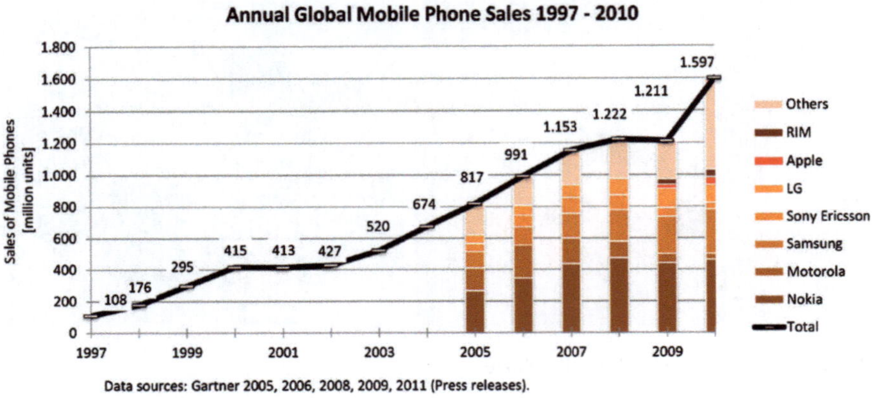

Fig. 6.7 Annual mobile phone sales 1997–2010

to nearly 1,600 Mio. sold mobile phones occurred. From 2005 onwards reasonable shares of the biggest manufacturers were available and have been added to the figure [7–11]. Estimates and prognosis concerning sales have generally been too low as e.g. Gartner estimated in 2005 a strong growth in sales with the prediction of 1 billion annual mobile phone sales in 2009. That already occurred nearly 3 years earlier between 2006 with 991 Mio. sales and 2007 with 1,153 Mio. sales [7].

6.3.4 Analysis of the Mobile Phone REPM

The REM EDX and RFA analysis was performed by a sample of nine speaker magnets, three earphones (in ear) and two headsets. The samples covered phones and systems from 2000 to 2008 as well as different manufacturers. The results are depicted in Fig. 6.8. Two samples, both earphones, contained no REE. Of the remaining samples all except of one showed between 15 and 28 % Nd (RFA) (between 30 and 50 % of Nd with REM EDX). The RFA analysis is preferable for the quantitative decision. It was noted that several samples could not be prepared well enough as the coating, usually Zn, could not be removed entirely so that the coating falsified the data. From a pre-test it was known that Zn is a primary constituent of some coatings. As the samples were very similar to the HDD samples an average for the mobile phone samples could be determined with 26 % of Nd and close to 5 % Pr. Five phone speaker samples and all earphones contained Pr, except the two containing no REE. For a more detailed insight, more samples would have to be analyzed.

Yet, the results allow the decision to refrain from further analysis—at least of old mobile phones–as the total amount of Nd required in 2010 was about 450 t

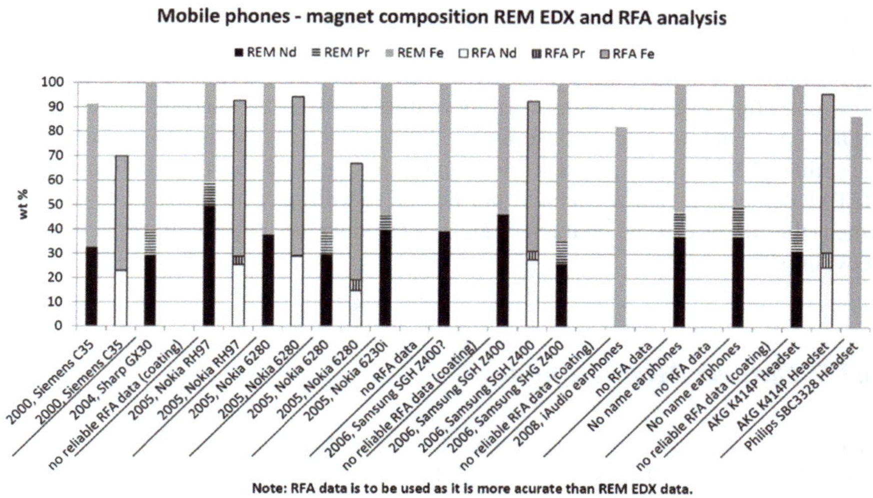

Fig. 6.8 Mobile phones–magnet composition REM EDX and RFA analysis

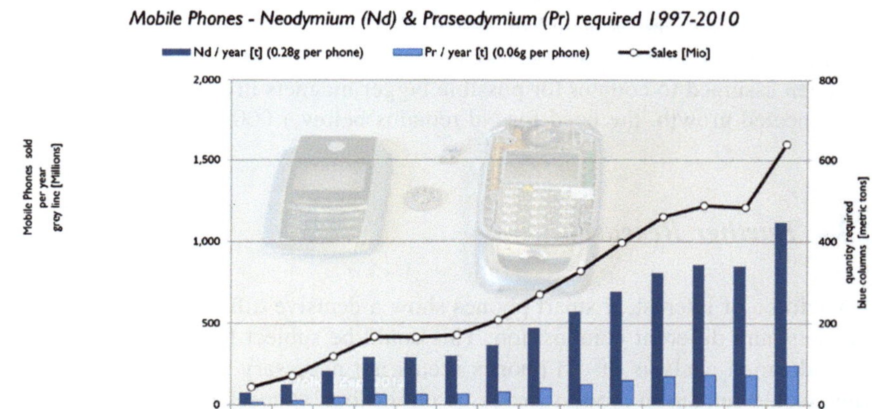

	1997	1998	1999	2000	2001	2002	2003	2004	2005	2006	2007	2008	2009	2010
Nd / year [t] (0.28g per phone)	30	49	83	116	116	120	146	189	229	277	323	342	339	447
Pr / year [t] (0.06g per phone)	6	11	18	25	25	26	31	40	49	59	69	73	73	96
Sales [Mio]	108	176	295	415	413	427	520	674	817	991	1,153	1,222	1,211	1,597

Data sources: Mobile Phones Sales - Gartner 2005 - 2011, Press releases. Nd, Pr required - own measurement and calculation.

Note: The average cumulative magnet weight per Mobile phone speaker, vibrating motor and earphones is 1.1g, thereof 26% Nd (0.28g) and 5% Pr (0.06g).

Fig. 6.9 Mobile phones–Nd and Pr required 1997–2010

when 1.6 billion mobile phones were sold (see Fig. 6.9). In total about 3,000 t Nd can be expected in all sold mobile phones from 1997 until 2010. These numbers show that the quantities needed for the production of mobile phone REPM are not a big share.

Compared to the 2007 production data of Nd with 21,141 t the 323 t Nd correspond to about 1.5 %; in 2010 with 22,256 t Nd annual production, the 450 t correspond to about 2 % share of the mobile phone sector in REPM.

A potential recycling seems even more difficult than the one for the magnets of HDDs. A total quantity of just under 3,000 t globally seems not efficient for recycling of REE out of phones. Here again the logistics and the small amount of REPM per phone sets a decisive hurdle for recycling. The matter may change, when the REE can be extracted as a byproduct of a general mobile phone recycling.

6.3.5 Outlook Mobile Phones

One disadvantage of the analysis was that no real iPhone or better *Smart phone* could be included. There is a potential chance, that these phones use slightly bigger magnets; however a big change in quantity should not be expected. So the question is how the global phone sales will develop and thus production and need for REPM. Llamas [12] forecasts 2.1 billion sales in 2015, this corresponds to a CAGR of 8.4 %. With this number the years in between 2010 and 2015 have been interpolated. They are depicted in Table 6.3.

In the table a higher quantity of 0.4 g Nd and 0.07 g Pr per mobile phone unit has been assumed to counter for possible bigger magnets in smart phones. Despite the expected growth, the need for Nd remains below 1,000 t per year until 2015.

6.3.6 Further Research

It would be of interest, if smart phones show a decisive different situation, bigger magnets, and different composition. This could be subject for further research. A more detailed analysis of old phones seems not necessary as the expected quantities would remain in modest shares. A closer look at speakers in general both in automotive and audio applications (home stereo, professional music equipment and TV) and at headphones would enhance the picture.

Table 6.3 Global mobile phone sales–forecast

	2010	2011	2012	2013	2014	2015
Global sales[a] [Mio.]	1,600	1,700	1,800	1,900	2,000	2,100
Nd required[b] [metric t]	640	680	720	760	800	840
Pr required[b] [metric t]	112	119	126	133	140	147

[a] 2010 and 2015 figures are forecasts from Llamas et al. [12]. Other data have been interpolated thereof

[b] Nd calculation based on 0.4 g Nd and 0.07 g Pr per phone

Mobile phones/speaker and vibrating motor magnet		
2005	230 t Nd	50 t Pr
2010	450 t Nd	100 t Pr
2015	840 t Nd	147 t Pr
2020	n/a	n/a

6.4 REPM Use in Wind Turbine Generators

One of the most challenging problems concerning REE needed in the wind industry is again the mostly inaccurate wording combined with an informational or knowledge deficit; at least in many reports. There is at the same time comprehensive information available in the scientific area about the different technologies and wind turbine systems (e.g. [13–17]). A detailed analysis gives e.g. the concept report of Polinder et al. [15, 16] as well as the WindPACT Study [18]. There generator topologies, mechanical and electrical concepts are discussed [15, 16]. In the media and abstracts usually headlines are posted stating that wind turbines need REEs (e.g. [19, p. 2, 20, p. 1, 21]). This however can imply that WTG cannot be produced without REEs, which is not at all true. In fact, the REE magnet based system is only one out of many others. This WTG design is usually coupled with the so called direct-drive (DD) system, but this can be manufactured with an electrically excited magnet (electrically excited synchronous generator EESG) or with a permanent magnet based synchronous generator (PMSG) [15, 16, p. 12f]. So the notion that any DD system uses a rare earth based permanent magnet is false. The global share of DD systems is given with 14 % (2009) and 17 % (2010) [22, 23]. Now these numbers combine both EESG and PMSG systems and may not be misunderstood, that 14/17 % of all Wind Turbine Generators (WTGs) are equipped with REE based technology. In fact about half of the DD WTGs are produced by the German company ENERCON; they use electrical excited magnets instead of REE based PM [24]. Another discrepancy arises from small but decisive differences in wording and reporting: usually global numbers are disseminated as MW installed (e.g. [23, 25, 26]), but Annual Reports talk about MW orders, sold, produced, shipped, delivered and installed: "In 2010, Vestas produced and shipped 2,025 wind turbines with an aggregate capacity of 4,057 MW, against 3,320 wind turbines and 6,131 MW in 2009. The decline was due to the low order intake in 2009. In total, 5,842 MW was handed over to the customers." [27, p. 9]. This 'handed over' number corresponds to the term 'installed' even though that is not entirely correct. The handover to the customer does not yet mean that the turbine has already been erected and connected to the grid, i.e. installed. But still there is room for another error as a WTG may be erected and a proper mechanical function be proofed without the WTG being connected to the grid. Another topic may also add to some inaccuracies: the accounting of fiscal year versus calendar year. The Suzlon group reports using fiscal year data for example [28, p. 104].

Eventually it is important to note that it has to be checked, which numbers are presented so that only comparable numbers are actually compared. For this work in fact, the production data would have been of highest interest when talking about REE needed in the wind industry business. The production data however were not attainable for all companies. In consideration and weighing of these factors it was decided to stick with installed data numbers. It will of course be kept in mind that a direct deduction of installed power to annual material purchase (including REE in some cases), is not entirely valid.

As an introduction to the topic Fig. 6.10 illustrates the global annual installed and cumulative wind power from 1996 to 2010. On top of the bars is a curve showing annual growth rates in percent to demonstrate the dynamics in this area. For the years 2008–2010, nearly invisible, permanent magnet relevant power capacities are displayed. This issue will be analyzed in the next paragraphs. It should become obvious that after a continuous growth rate from 2004 to 2009 a sharp decline in growth occurred. Still the annual global installed power varies slightly around 40 GW, so that in 2010 a global cumulative total of nearly 200 GW was reached. The outlook however foresees an increasing demand.

The following paragraphs try to assemble comparable data and detail the topic of the different types of wind turbines before a description of the top 15 wind turbine manufacturers will follow. After that a summary of the production i.e.

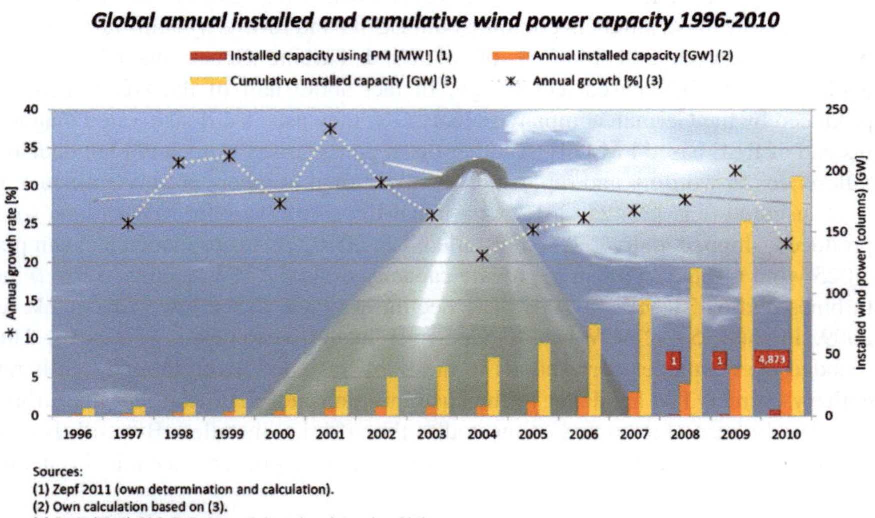

Fig. 6.10 Global annual installed and cumulative wind power capacity 1996–2010

installed power capacity is given with a calculation of actual ‘installed’ REEs. Finally an outlook gives a hint of what can be expected concerning REEs in the wind industry in near future.

6.4.1 Data Issues WTGs

Accurate data about quantities of REE used is given as estimates only in several reports. Only a few scientific papers deal with the amount of Nd required or magnet weight necessary per MW of installed power. For global sales and ‘installed’ numbers mostly annual reports have been used.

6.4.2 Types of Wind Turbines

There are both detailed scientific and industrial information available about wind turbine design, construction and operation so that here only a very short and simplified description is given. Again it is important to pay attention to small differences in wording and definitions to understand the REE relevant aspects around the wind energy. More detailed information is given e.g. by Heier [13], Gasch and Twele [29] or the Wind Energy Technology and Planning information page accessible via the homepage of the World Wind Energy Association (WWEA www.wwindea.org). These authors’ information has mainly been used to assemble the following short summary.

In principal a wind turbines task nowadays is the production of energy. In more detail it can be said that the turbine converts kinetic energy via the transformation from mechanical into electrical energy. To accomplish this, first a wind energy transformer is needed: the wind turbine itself with its rotor blades. Wind then turns the rotor when in a next step the revolutions of the rotor have to be transformed in speed and torque by means of a gearbox to feed the mechanical energy into a generator. This generator converts the mechanical energy into the required electrical energy, depending on the grid features, e.g. into 50 Hz power. Before the electric energy is then fed into the grid, some protection and conversion units may be interconnected to guarantee a smooth grid adaption. The classical conversion systems are based on asynchronous generators (ASG) and require all the before mentioned components. The mechanical drive train and the gearbox consist of many parts that carry strong mechanical loads and are thus susceptible to wear and tear. This technology is said to be more prone for failure than the direct drive topology. The asynchronous technology is still the one used in most of today’s WTGs.

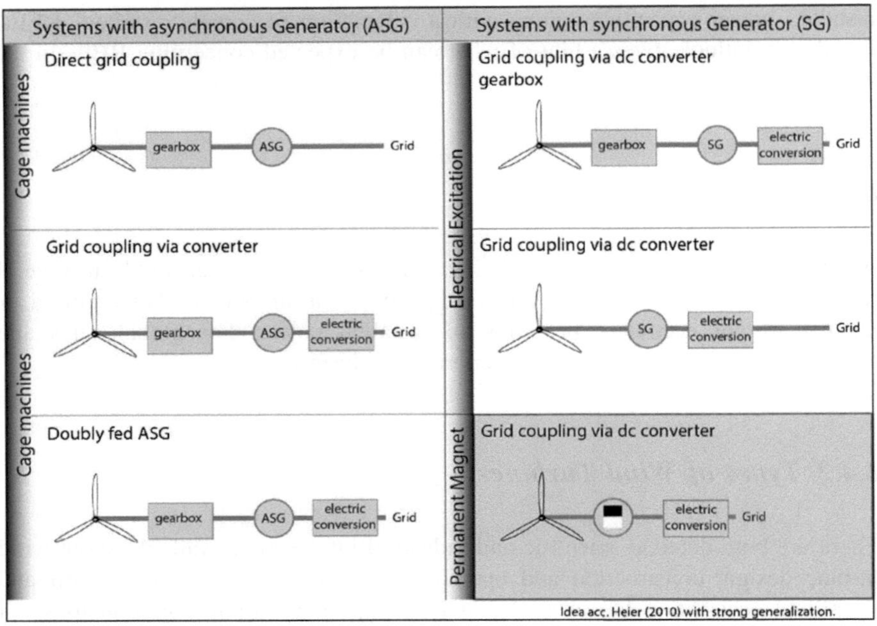

Fig. 6.11 WTG system designs

An alternative to this classical approach is the use of synchronous generator (SG) systems. These SG again appear in versions with gearboxes between the rotor and the SG; another design is based on permanent magnets in a direct drive system which does not require a gearbox. There are again several basic principles to implement the direct drive topology. In particular separately (electrically) excited and permanently (REE) excited versions come into consideration [13, 29]. Figure 6.11 gives a quick and approximate overview of standard WTG designs.

Prominent examples for the separately excited design principle are the present WTGs of the German company ENERCON. They use a direct drive annular generator with conventional copper windings. The annular generator presents a low speed synchronous generator with no direct grid coupling which is accomplished by a separate DC link and an inverter. The stator is manufactured with conventional copper windings which is done manually within the company to guarantee the desired quality. The magnetic field of the stator then is excited via pole shoes that are located on the rotor, i.e. the moveable part inside the generator—not to be confused with the big rotor (blades) [30]. Thus ENERCON uses the direct drive system but does not use REE based permanent magnets. This often leads to misunderstandings.

In all direct drive designs the stators are usually almost identical but the rotor design is different. As has been said does ENERCON use a wound field system with pole shoes as exciting array, whereas the rotor of the permanent magnet design uses REE based permanent magnets at the rotor. Due to the principally simple design with this REE based permanent magnet generator (PMG)[3] lots of poles can be manufactured which eases the adaption to slow rotation speeds [31], i.e. low wind situations. In comparison to the wound field rotors which need to be powered, the permanent magnet technology does not need any other excitation power. This results in about a third less excitation losses and eventually in a higher efficiency. Despite the necessity of a subsequent power converter and transformer in REPMGs [13] this design is lighter in weight than the electrical excitation approach. Rollik and Schleede [32] state a possible weight reduction of 70 t for one ENERCON E112 WTG with a 124 m high rotor axis and blades with a diameter of 114 m [32].

One advantage of the REPMGs is that first there is no need for a gearbox and second that the simplicity of the material does neither require further maintenance nor does it require additional parts so that during the entire life time less maintenance is expected [31, 33]. This is especially important for offshore applications as maintenance can get critical in adverse weather with storms, turbulences and high waves, so that access to malfunctioning WTGs is not guaranteed. The plans for offshore installations grow steadily so that more effort in the evolution of the permanent magnet design can be seen. Interestingly there have been studies performed aimed at the analysis of this advantage; however they came to the conclusion that no increased reliability has been exhibited (as stated in [34]). One more problem of the direct drive systems is the general bigger weight of the nacelle compared to gearbox systems. Shrestha et al. [17] give detailed insights in estimated weights of WTGs ranging from 1 MW up to hypothetical 20 MW designs.

A further approach design is a hybrid system using a combination of the permanent magnet based array and a single or a two-stage gearbox system which is used e.g. by the Multibrid wind turbine concept of the French Areva company. Several studies around an optimum drive train resulted in this hybrid approach as the best solution with still some areas for improvement left; for details refer to [33]. This hybrid drive train solution is said to have the lowest weight compared to asynchronous and pure PMG designs [32] because both the PMG and the gearbox array can be built smaller than in the pure versions. Heier [35] states 320 t for a Multibrid 5 MW system and 350 t for a RePower 5 MW system [35, pp. 23, 33]. This hybrid approach is favored by a number of companies including Winwind, Innovative Windpower and Gamesa [33], which is the only company of the mentioned ones ranking under the top 15 in 2010.

Finally there remains the question about how much REE are needed for a REPMG? Data is not obviously attainable and published numbers and estimations differ. Jensen [14] states for standard REPMG systems about 0.5 t of permanent magnet per MW of installed power. Lifton [36] attributes between 0.7 and 1 t PM per MW, Oakdene Hollins [20] calculates with an average of 700 kg PM per MW

---

[3] REE based PMG is abbreviated ***REPMG*** as the wording DD-system or PMG alone could cause misunderstandings.

Table 6.4 Magnet weight in wind turbines

Magnet weight in wind turbines—a comparison					
	Jensen [14]	Lifton [36]	Oakdene Hollins [20]	Polinder et al. [15]	Average (Jensen and Polinder)
PM/MW [kg]	500	700–1,000	700	600	550
Nd/MW [kg]	150	210–300	210	180	165

Notes
Calculation of Nd required based on the assumption, that about 30 % of a $Nd_2Fe_{14}B$ PM is Nd
PM permanent magnet; *MW* megawatt

Table 6.5 Magnet weight in wind turbines (summary)

Magnet weight in wind turbines						
WTG size	1 MW	1.5 MW	2 MW	3.6 MW	5 MW	7 MW
Weight of PM [kg]	550	825	1,100	1,980	2,750	3,850
Weight of Nd [kg]	165	248	330	594	825	1,155

Notes
Calculation based on an average value derived from Jensen [14] and Polinder et al. [15]

installed and new design approaches for 10 MW WTGs list 6 t of PM for a 10 MW turbine [15] (Table 6.4).

In this case Jensen and Polinder et al. as scientists, are given more credibility than Lifton and Oakdene Hollins for reliable numbers so that as an average 550 kg of PM or 165 kg per MW of Nd is assumed. Table 6.5 shows a simple calculation of probable needed magnet weight and the Nd required–based on a share of 30 % of Nd.

The actually used quantities could not be determined. It is also not known in how far in hybrid systems the magnet weight is less than discussed here. This definitely remains an important subject for further research.

Upcoming designs and research is done around both the increase of power of a turbine and the substitution of REPMG. The size issue centers around 10 MW turbines where several companies are involved. The praised Britannia project however, which was announced in 2009 was put on hold in August 2011 with no detailed reasons given [37]. Research is yet ongoing with results and timescales not yet foreseeable. Further research centers around the substitution of REE, e.g. by means of induction generators, as [38] suggests. Another idea is the implementation of high temperature superconducting (HTS) materials as reported by Windpower [33], Hopwood [39] or AMSC [40]. For all of these technologies no market entries are visible yet.

This short introduction shows that only a small range of turbine technology is based on REE containing permanent magnet technology. In this context the headlines raising concerns that wind turbine construction is endangered by a shortage of REE is not actually correct. There are alternatives to the REE based turbines, of course not considering any possible better efficiency on the material or economic side. This topic seems to be balanced from a factual point of view, however as Hopwood [41] mentions, "it appeared that Direct Drive was winning the propaganda battle, until Vestas announced its new V164 machine, the first in the company's history specifically designed for offshore wind. One of the biggest surprises… was that Vestas bucked the direct drive hype to go for a geared solution." Interestingly enough does the company homepage and the brochures of the new V164 praise the new concept which requires less maintenance and will use a permanent magnet generator! [42].

6.4.3 Wind Turbine Manufacturers

According to BTM Consult, the top 10 wind energy producers in 2010 accounted for about 80 % of the global share of wind energy. For numbers 11–15 no detailed data was available so that their share had to be calculated using some logical assumptions. The Chinese company XEMC (ranking at position 15) already delivers REE based WTGs. Calculating back from data out of BTMs Consult report [43], XEMX provides a share of 0.9 % to the annual installed capacity. If XEMC at position 15 has 0.9 %, positions 11–4 must have shares of at least 0.9 % and at the most 4.2 % as that value is given for position 10. If calculated very conservatively these 5 companies provide at least 5 %, probably around 8 % to the installed capacity so that the top 15 companies account for 85–88 % of the annual installed capacity.

According to BTM Consult [43] the top 15 companies in 2010 were:

Wind turbine manufacturers 2010, Top 15		
1. Vestas, DK	6. Gamesa, ES	11. Mingyang, CN
2. Sinovel, CN	7. Dongfang, CN	12. Nordex, GE
3. GE Energy, US	8. Suzlon Group, IN	13. Mitsubishi, US/JP
4. ENERCON,GE	9. Siemens, DK	14. Sewind,CN
5. Goldwind, CN	10. United Power, CN	15. XEMX, CN

This ranking is only a snapshot as the wind turbine market is volatile. Especially in China new companies press for mass production of MW scale turbines as well as several of these top 15 companies strive for 5–6 MW offshore turbines. In parallel there is research and development ongoing both to optimize existing technologies, to look for substitutes for REE based turbine designs as well as the search for even more modern approaches like hybrid systems or superconducting materials. The latest turbine size which is under development is in the 10 MW class.

These top 15 companies will be described briefly to show their portfolio and the scales in which they perform. At the end of this description, a table will summarize all the data given. The data is based on 2010 figures and was mostly taken from the companies' annual reports.

Vestas is presently the top ranking company. At the end of 2010 Vestas has delivered 43,433 WTGs which have an installed capacity of 44,114 MW power. Until 2010 Vestas only produced classical asynchronous wind energy generating technology. In 2010 the company announced a whole range of new WTGs, partially named with the appendix Gridstreamer^{TM}, using direct drive technology based on REE permanent magnets. In 2010 the first two V112-3.0 MW turbines have been installed producing a total of 6 MW power [27, p. 18]. From 2011 onwards the new technology is offered for all wind and turbulence classes from 1.8 MW up to 7.0 MW turbines [27]. Interestingly however is to note that Vestas' CEO Engels mentioned in an interview done in January 2011 that their quest is to give the customer the best cost of energy. He argued that the direct drive technology still being unproven technology and that gearboxed technology has a better performance than their reputation would suggest so that in this context the gearbox would be the best solution [41]. In January 2011 as well Vestas issued the first brochure of the V112-3.0 MW turbine using direct drive technology [44] and in early February 2011 a press release announced the presentation of the new 6 MW turbine in March of 2011 [45]. In October 2011 another press release stated that orders for the new V112 direct drive turbine reached the milestone of 1 GW [46]. Compared to the 4GW produced total in 2010 this represents an impressive number, even though the production, shipment and installation will extend to further than 2012.

Sinovel, the Chinese *Sinovel Wind Group Co., Ltd.* is presently the second largest producer of wind turbines globally and the biggest in China. The company delivered its first 1.5 MW turbine in June 2006 and until the end of that year already 100 turbines had been produced. The main product is the SL1500 turbine rated with 1.5 MW using asynchronous technology. In 2007 already 750 MW (500WTGs) have been produced growing to 1,403 MW (935 WTGs) in 2008, 3,510 MW (2,317 WTGs—2,294 of the 1.5 MW type and 23 of the 3 MW type) in 2009 to reach 4,386 MW with 2,903 WTGS (2,882 WTGs of the 1.5 MW class and 21 WTGs of the 3 MW class). After in 2009 the first 3 MW offshore turbine was installed to equip the 100 MW Shanghai Donghai Bridge Demonstration Offshore Wind Power Project, the first 5 MW turbine was launched in October 2010 until in May 2011 the first 6 MW turbine was introduced. All of them work with asynchronous technology [47]. No REE based permanent magnets are used.

GE Wind, a business of the General Electric Company, produces wind turbines since 1996, when the first 1.5 MW turbine was installed. Until 2011 about 15,000 WTGs have been shipped. The standard technology is the classical asynchronous drive until in 2005 an offshore test-installation with 13 turbines with a direct drive technology was set up [48]. This installation was used as a test bed to develop new turbines based on REPM Direct Drive technology: the 4.0-110 turbine was introduced in 2010 and the slightly stronger 4.1-113 system in 2011. No information on actual installations could be gathered.

Goldwind, actually *Xinjiang Goldwind Science & Technology Co., Ltd.*, was established in 1998. Their main products are a 1.5 MW and a 2.5 MW turbine, both of which are DD systems based on REPM. The 1.5 MW turbine entered service in 2007. In 2009 the bigger WTG with 2.5 MW, mainly for offshore use, was introduced and in 2010 the first turbine was installed. However, the 1.5 MW turbine is produced in considerable quantities: in 2007, about 300 MW were installed, in 2008 already 727.5 MW (485 WTGs); in 2009 about 1,590 MW (1,061 WTGs) and in 2010 about 3,850 MW (2,567 WTGs) were installed and 4,007 MW sold. Next to these two main products a new 6 MW turbine is under development [49]. BTM Consult [23, 43] shows for Goldwind a global share of 9.5 % or 3,743 MW [43] which shows a slight difference to the numbers given in the Annual report by Goldwind itself. Yet, Goldwind ranks around number 5 and 6 of the top producers and is the biggest company using REPM technology.

ENERCON, the biggest producer of Germany was founded in 1984 and initially produced gearboxed wind turbines. In 1992 the company switched to gearless technology and enhanced that technology until today. ENERCON is one of the leading companies worldwide and installed more than 16,000 wind turbines in over 30 countries. "The drive system for ENERCON wind energy converters is based on a simple principle: fewer rotating components reduce mechanical stress while at the same time increasing the equipment's technical service life." [30, p. 11]. The system is composed of the stator with copper wirings and the magnet field of the stator is then excited by the rotor with a special array of pole shoes. There are no REE used for the magnets in the annular generator [24, 30].

As ENERCONs turbines are of the Direct Drive technology, their WTGs are often misinterpreted as REE based systems, which is not the case. ENERCON, Goldwind (and XEMX) both use DD systems. Usually they are considered as one technology group which represent 17.6 % of the world market [43]. This bears the misconception, that the global share of REE based systems is again this 17.6 % which is not the case. The ENERCON share has to be subtracted from the 17.6 % so that the actual global share of REE based systems is about 10 %.

Gamesa is a Spanish producer ranking around position 8. The company was founded in 1976 and began with wind energy activities from 1994 onwards. At the end of 2010, Gamesa has installed a total of 20,834 MW. The main products are the G5X-850 kW series (24 % of sales in 2010), the G9x-2.0 MW series (71 % of sales) and the subsidiary companies MADE, accounting for 5 % [50]. In 2009 Gamesa started to introduce the new G10X-4.5 MW turbine for onshore, low to medium wind use, which is based on a hybrid REE permanent magnet synchronous generator design in combination with a 2 stage gearbox. Presently there are two versions available, the G128-4.5 MW turbine and the G136-4.5 MW turbine [50, 51, 52]. A further agreement has been set up with E.ON for delivery of G10X turbines for an offshore wind park in Italy and the installation of a prototype in 2011 [50, p. 153]. For the offshore arena further new turbines are planned, the G11X-5 MW and the G14X-6-7 MW turbines. Together within a strategic alliance with Northrop–Grumman Shipbuilding the G11X-5.0 MW will be developed with the intent to set up a first demonstration site end of 2012. The stronger G14X turbine is scheduled for 2014–2015 [53]. So at the end of 2010 two WTGS with this REE permanent magnet technology were fabricated. The future trend however is not sure.

Dongfang, the Chinese Dongfang Electric Corporation (DEC) group produces around 30,000 MW of power generating in various fields including hydro, thermal nuclear and wind. As such, Dongfang started in 2004 in the wind energy business using RePower technology and is now the third largest Chinese wind energy producer and ranks around number 6 globally. Information about the company and production data is scarce. According to AMSC [53] the main product was a 1.5 MW turbine of which 800 were shipped in 2008. The company states at their homepage an annual output of 500–800 units including 1.5, 2.5 and 3.0 MW turbines [54]. In the same message a hint is given on the development of a direct drive system whereas no advice on the use of PMG or other material is given. It is assumed however, that the company develops a PMG which is also underlined by the cooperation with the AMSC. AMSC are promoting REPMG and HTS systems so that it is likely that in the cooperation with Dongfang, REPMG will appear in near future. Until 2010 it is believed that the production was entirely based on classic asynchronous systems. According to data published by BTM (2011) Dongfang produced a cumulative total of about 3,640 MW with more than 2,000 WTGs. In 2009 GWEC states 2,475 MW of installed power [25, p. 11].

Suzlon Group and RePower are considered together as Suzlon factually owns RePower and uses the same technology. Together the two companies rank around position 8. The installed capacity in 2010 was 2,719 MW [43] whereas taken from the company's annual reports a total of 2,009 MW has been extracted [27, p. 73, 55, p. 100]. The technology used by both companies is of the asynchronous type. Turbine capacities range from 600 kW up to 6 MW.

Siemens AG and Siemens Wind Power A/S are in the wind industry business since 30 years and have turbines installed both onshore and offshore. The first offshore wind park was installed 1991 in Vindeby, DK Baltic Sea, with an installed capacity of 4.95 MW provided by 11 turbines rated with 450 kW each. As of June 2009 Siemens had 642 MW of installed offshore capacity with nine projects in progress aiming at around 2.5 GW of installed power using still asynchronous technology [56]. Siemens today offers a portfolio of asynchronous machines rated with 2.3 MW and 3.6 MW and three new turbine designs based on REPMG: the SWT 2.3-113 (for moderate winds), SWT 3.0-101 (strong winds) and SWT 6.0-154 (special offshore design for strong winds) (all information taken from [57]). The installed capacity in 2010 was 2,325 MW.

United Power of China started with the wind business in late 2006 after a technology transfer contract with the German Aerodyn Co. and installed three 1.5 MW prototype turbines in December 2007. In 2008 already 100 turbines were produced. In 2009 the company homepage says that the 800 annual production capacity plan was reached which is most probably meant as MW value, i.e. about 530 WTGs. United Power is grouped into category II of Chinese wind turbine manufacturers which are supposed to have the capability for mass production of MW-scale turbines and for 2009 an installed power value of 768 MW is given [58, p. 35f]. This as well corresponds to the assumption that the number 800 stands for MW production capacity instead of WTGs produced. In 2010 the 2,000 capacity plan (ca. 1,300 WTGs) is reported together with the delivery of a 3 MW prototype and the reception of a first overseas order (all data taken from United Power Homepage [59] except when cited separate). The technology used is of the asynchronous type whereas a 6 MW turbine is under development with REPMG technology. According to BTM (2011) United Power ranks number 10 globally and accounted for 4.2 % of the 2010 installed power (39.4 GW) so that a capacity of 1,655 MW results. This comes at least in the vicinity of the company homepages data. These discrepancies illustrate the problem associated with published data of different sources which all should be considered as reliable, both the company homepage as well as the consultants and global wind energy analysts. A more detailed insight could be gained when purchasing the consultant report, however there remains doubt whether these data can be verified. Nevertheless the relative quantitative direction seems clear and obvious. Doubts about the Chinese reporting remain which announce that *the plan has been fulfilled*. This has a scent of reporting for the fulfillment of the plans sake and not because the requirements have actually been achieved.

The Chinese Ming Yang Wind Power Group Limited (***Mingyang***) was founded in June 2006 and produced its first turbine in August 2007 [60]. Mingyang grew to the fifth largest Chinese producer in 2009. Together with United Power and XEMX, Mingyang is attributed to the category II Chinese wind companies, capable of mass producing MW-scale turbines. Their main product is a 1.5 MW scale turbine which is offered for the Chinese domestic market with various adaptions to environmental influences like climate and dust. In 2009 Mingyang installed 748.5 MW of power [58, p. 35]. In cooperation with the German aerodyn Energiesysteme GmbH and the aerodyn Asia Co. Ltd. an advanced super compact drive (SCD) is being developed for installation in various new turbine designs of 2.5/3.0 MW (onshore) and 5/6 MW capacity (offshore). This SCD-technology is in principal a hybrid system as has been described above using a 2-stage gearbox and a subsequent REPMG. This allows a reduction in size and weight and the elimination of the biggest disadvantages of the asynchronous and pure PMG designs [61]. In 2010 the first 3 MW SCD turbine prototype was installed. Information about further orders is not known.

Nordex is a German company that has its main competence in a 2.5 MW turbine which is produced since 2000. The technology is based on asynchronous technology and is offered for all wind classes in various layouts [62]. A new 6 MW turbine as PMG is announced for offshore installations [63]. In 2009 about 1,060 MW, in 2010 about 910 MW were installed [64].

Mitshubishi ranks at position 13 and is using asynchronous technology on turbines rated 1 MW and 2.3/2.4 MW [65]. On the other hand there are reports that Mitshubishi is using direct drive systems with permanent magnets, however it is not stated whether that is RE based (Bywaters et al. [18]).

Sewind is a joint venture of the Shanghai Electric Group Co. Ltd. (SEC) and Huadian Engineering Co., Ltd. which was established in 2006. In 2007 the first WTG was produced with imported technology from the EU Energy Wind Group Corp. Ltd., UK, and in cooperation with the German aerodyn. The main products are turbines with 1.25 and 2.0 MW (offshore) capacity, which entered a test run in 2009 [66]. A new 3.6 MW turbine was developed and a prototype installed in mid-2010. Sewind globally ranks at position 14 and is, next to Sinovel, Goldwind and New United the only Chinese company which exported a small number of turbines [58]. The technology used is not known but assumed to be of the asynchronous type. The new 3.6 MW turbine might be of the REPMG type.

XEMC, Xiangtan Electric Manufacturing Co., Ltd. is one of the largest manufacturing complexes with an emphasis on generator production of all kinds. Amongst the generator group a wind sector has been set up which ranks at number 15 globally. Cooperation with the Dutch Darwind has been set up. The main product of XEMC is a 2 MW REPMG turbine with a 5 MW REPMG offshore turbine being under development. In late 2010 the first 5 MW PMG was unveiled in Xiangtan [67]. As no direct production data was available, the probable production was estimated using data from [43] which state that 17.6 % of the world market account for the direct drive design. Subtracting Goldwind (9.5 %) and ENERCON (7.2 %) leaves 0.9 % for XEMC which is equivalent to 355 MW of installed power.

Clipper Windpower is a United Technologies Corp. company and claims a 25 years of experience in wind power business. The main product is the 2.5 MW Liberty turbine based on a REPMG system. The prototype Liberty turbine was installed in 2005 and in 2007 the serial production began. Today Clipper manages over 6,500 MW of windpower. It is not said, whether this is all from own turbines however [68, 69]. A new 10 MW turbine was under development, the Britannia turbine, however the project has been put on halt [38].

There are more companies working on 5 up to 10 MW designs mainly planned for the implementation of PMG technology [70].

6.4.4 REPM Use in WTGs

Table 6.6 shows the installed capacities along the top 15 WTG manufacturers for the years 2009 and 2010. It can be seen that for several companies no company data could be derived instead GWEC or BTM consult data had to be used. It can also be seen that some data discrepancies between company and GWEC or BTM exist, e.g. the Goldwind company data for 2009 states less installed power than GWEC reports; for 2010 the company shows about 250 MW more than BTM reports. The message from this table is another one, i.e. the actual REPMG based WTGs which are marked with a grey background color. For 2009 an installed power for REPMG based systems with 2,045 MW total could be aggregated and for 2010 there were already 4,223 MW installed. This represents about 11 % of

Table 6.6 Overview of annual installed wind power capacity, 2009 and 2010

Overview of annual installed wind power capacity 2009 and 2010.
Various sources with varying information.

Rank[1]	Manufacturer	System[2]	2009 installed capacity: Company data [MW][3]	ref	GWEC [MW][4]	2010 installed capacity: Company data [MW][3]	ref	BTM [MW][1]	BTM [%][1]
1	Vestas, DK	REPM	none	ve	4,766	**6**	ve	5,832	14.8
		other	4,764	ve		5,836	ve		
2	Sinovel, CN	REPM	none	si	3,510	none		4,374	11.1
		other	3,510	si		4,323	si		
3	GE Energy, US	REPM	none	ge	4,741	n/a		3,783	9.6
		other	n/a			n/a			
4	Goldwind, CN	REPM	**1,591.5**	gw	2,727	**3,853**	gw	3,743	9.5
		other	444	gw		154	gw		
5	ENERCON, GE	REPM	none	en	3,221	none	en	2,837	7.2
		other	n/a			n/a			
6	Suzlon Group, IN	REPM	none	sr	3,718	none	sr	2,719	6.9
		other	1,828.2	sr		2,009	sr		
7	Dongfang, CN	REPM	none	cw	2,475	none	3	2,640	6.7
		other	n/a			n/a			
8	Gamesa, ES	REPM	none	ga	2,546	**9**	ga	2,601	6.6
		other	2,231	ga		2,577	ga		
9	Siemens, DK	REPM	n/a		2,265	n/a		2,325	5.9
		other	n/a			n/a			
10	United Power, CN	REPM	none	cw		n/a		1,655	4.2
		other	768	cw		n/a			
11	Mingyang, CN	REPM	none	cw		none	my		
		other	748,5	cw		n/a			
12	Nordex, GE	REPM	none	no		none	no		
		other	1,059.5	no		909,2	no		
13	Mitsubishi, US	REPM	none	mi		none	mi		
		other	n/a			n/a			
14	Sewind, CN	REPM	none	sw		n/a			
		other	n/a			n/a			
15	XEMX, CN	REPM	**454**	cw		**355**	1,c		
		other	n/a			n/a			
	Other Manufacturers				7,034			8,235	20.9
	Annual Total				37,003			40,744	103.4
	Total REPM		**2,045.5**			**4223**		~11% of annual installed capacity	

(continued)

Table 6.6 (continued)

Notes
Grey shading highlights actually installed PM capacity
[1] BTM Consult [43]. Data given in a pie chart in BTM Consult [43] add up to 103.4 %, thus the sum of MW installed adds up to 40,744 MW, which is 103.4 % of the given/actual 39,404 MW installed
[2] System: *PM* permanent magnet based; other–asynchronous design or direct drive without permanent magnet
[3] Company data were only given when such were available. They usually differ from the other reports; partially substantial
[4] GWEC [25]
n/a not available
c calculated
cw Junfeng et al. [58]
en ENERCON [24]
ga Gamesa [7, 51, 52]. Annual report 2010, p. 185. (2009 number calculated from data p. 6)
ge GE [48]
gw Goldwind [49]
mi Mitsubishi [65]
no Nordex [64, p. 47]
si Sinovel [47]
sm Siemens [57]
sr Suzlon [28]. Annual report 2010–2011, p. 100. RePower [55]. Annual report 2010-2011, p. 2
ve Vestas [27, p. 17f]

the considered wind power. As these top 15 companies account for 85–88 % of the global annual production, the share of REPM for the remaining 15–12 % WTG manufacturers not described above have to be included. Here again a share of 11 % of REPMG systems within these remaining producers is assumed. To stay on the conservative side, 15 % remaining manufacturers are taken from 39.4 GW which equals 5.91 GW of not covered wind power by the top 15. Thereof 11 % assumed using REPMG results in 0.65 GW or 650 MW, so that finally a total of 4,873 MW of REE based wind generation in 2010 can be taken as a basis. This converts back to 12.4 % of annual total installed wind power. **As a last step the 4,873 MW are multiplied by 165 kg of Nd to get the total amount of Nd used: 804 t**. Using the same idea for 2009 where 2,045 MW of REE based wind power was identified for the top 15 producers, an amount of 2,270 MW total is derived resulting in 375 t of Nd required.

6.4.5 Outlook

Figure 6.12 shows the growth scenarios disseminated by the Global Wind Energy Council in cooperation with Greenpeace International and the German Aerospace Centre (DLR). Three scenarios were elaborated starting from a reference scenario via a moderate to an advanced scenario. The reference scenario assumes

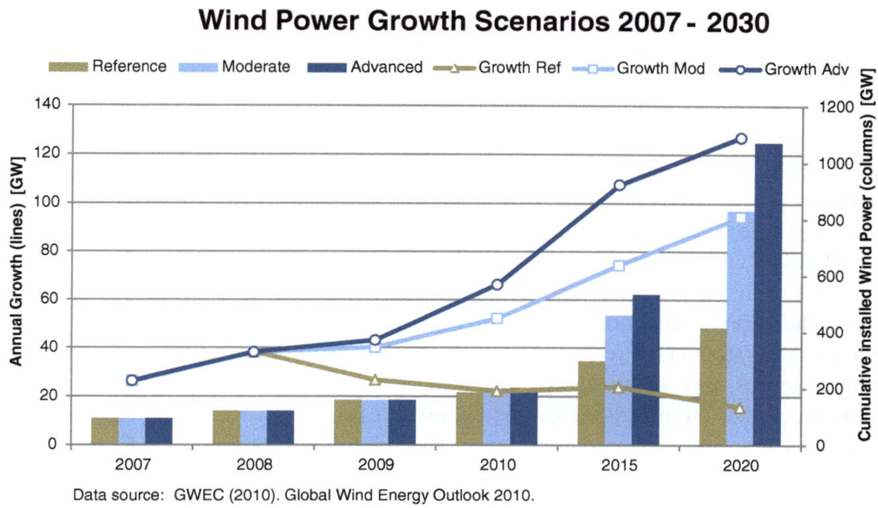

Fig. 6.12 Wind power growth scenarios 2007–2030

conservatism and takes into account only existing policies and measures. Using these assumptions the annual growth is actually shrinking from around 40 GW in 2008 to a little less than 20 GW in 2030. The cumulative total then would be around 400 GW in 2030. The moderate scenario expects that all policy measures in support of renewable energy and carbon emission reduction will be implemented so that a continuous increasing annual growth will result reaching the 100 GW mark shortly after 2030. The cumulative total installed wind power then would be around 800 GW. The most ambitious scenario is the advanced one, where an unambiguous commitment is assumed which would result in a strong growing annual growth reaching about 65 GW in 2010 until 130 GW of installed power would be the annual growth in 2030 [71]. As a comparison, this equals the cumulative total in 2009.

To get back to the topic of actual and potential REE use in the wind industry, the problem is manifold. First the question is on how the wind market will develop. The growth scenarios show quite different growth rates and market development is not at all foreseeable. In 2010 the annual growth was 35.8 GW [25, p. 12] which is somewhere between the reference and the moderate scenario. Order books of the top 15 manufacturers hint to a stronger growth within the next few years, but here again the order-production-installation topic is an issue of uncertainty. Next comes of course the question on how strong the REPMG design will establish and change the market. It is very likely that the market for REPMG will grow strongly as manufacturers which had and have their portfolio based on asynchronous design change their product line in favor of REPMG designs and even in favor of hybrid or designs without REE. Even though studies suggest that the question of reduced maintenance and thus more overall reliability and efficiency is even, this reduced maintenance argument is the main one given to promote REPMG designs, disregarding e.g. the higher cost and weight of the gondola. But also the higher efficiencies are put into the field and should be kept in mind for an overall comparison. Nevertheless the market entry of big companies like Vestas, GE Wind, Gamesa, Nordex and others offers an argument that the amount of REPMG turbines will increase significantly in foreseeable future. As well there is evidence that hybrid designs gain attention and market share. It is also likely that by further research and optimization the required amount of REE per MW may be reduced. The near future, i.e. the next 2–3 years will tell how much the REPMG technology will grow and whether assumed REE shortages will act as show stoppers. Predictions and scenario development seem to be inappropriate yet as too many issues are still open. It is remembered that Vestas announced that the milestone of one Gigawatt of orders of REPMG based turbines has been reached. Other major players like Goldwind, Siemens or Gamesa also report increasing orders. Chinese companies are also expected to continue work on 5–6 MW class designs (as is prescribed in the present five-year-plan (see Chap. 3) with the REPMG technology being first choice. Eventually it can be said, that the amount of REPMG turbine installations will most probably increase considerably. But, recent headlines indicate that the offshore wind power yet inherits some major problems, mainly concerning the grid connection.

At last there is the challenge to predict a share for REPM WTGs in 2020. As no reliable data set was available, I assumed a share of 30 % for these systems. In the moderate scenario in 2020 an annual growth of 95 GW is expected. 30 % thereof or 31.5 GW would result in about 5,200 t Nd required for this scenario. This consideration of course does not take into account any improvements in design or substitution of Nd partially by Pr. But this assumed growth, 31.5 GW in 2020 for REPMG, represents todays' total annual installation rate which should give an idea of the relative share in this arena.

Wind energy|REPM based	
2005	n/a probably below 100 t
2010	804 t Nd
2020*	5,200 t Nd

* Assuming the moderate scenario with an annual growth in 2020 of 94.54 GW and a share of 30 % REPM based WTG (31,500 MW)

6.4.6 Further Research

Obviously the data reliability and validity is not given for all manufacturers. Better data about the manufacturers and their used technologies are required to determine real growth in REPMGs. And of course the actual composition of the magnets is necessary to actually determine the required amounts of the REE for this application field. One more thought could be some life-cycle or lease discussions with manufacturers about possible redemption scenarios for end-of-life systems and their re-manufacture or re-cycle.

6.5 REPM Use in Electro Mobility

In this chapter only e-mobility of automobiles is briefly analyzed. The big market of e-bikes is neglected even though there considerable quantities of produced e-bikes are expected. For 2010 about 31 Mio e-bike sales globally are reported [72, 73]. If an average magnet weight of 100 g would be assumed; i.e. about 30 g of Nd per e-bike, the global sales in 2010 would account for around 930 t Nd.

Figure 6.13 gives a qualitative overview of REE which are used and built into todays' cars. It is important to note, that most of the applications are built into todays' conventional cars already in considerable numbers; which are not obvious however. For the new envisaged e-mobility the electric traction motors and the battery or energy storage systems will be added on top of the already used REE (grey areas). This should be the important message that the new e-mobility does not invent the REE use in mobility but enhance especially the magnets and storage use. The fact that so many applications are built into the cars today already could turn into a decisive factor concerning criticality. This issue is not in the scope of this work so that it is not pursued further.

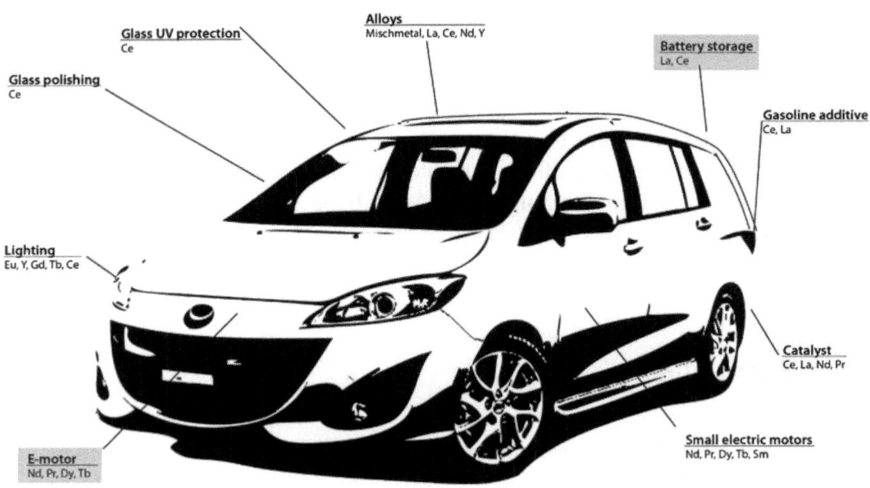

Fig. 6.13 REE use in cars

6.5.1 Data Issues E-Mobility

The topic of REPM use in electro mobility has just recently been thoroughly investigated and published by the German Institute for Applied Ecology (Öko-Institut e.V.) with the title 'Ressourceneffizienz und ressourcenpolitische Aspekte des Systems Elektromobilität' (in English: Resource efficiency and resource-political aspects of the system of electro mobility). Under the lead of the Institute companies like Daimler AG, Umicore and the University of Clausthal did a joint research on the ecological aspects of electro mobility. Thereby the issues of raw materials use have been discussed in depth. Several scenarios have been analyzed and calculated so that a range of possible demands driven by electro mobility could be derived. Detailed information is recommended to be obtained from the study itself [74], so that here only some key points shall be discussed.

One of the most discussed issues is the need of strong permanent magnets for electric and hybrid vehicles. In general an automotive hybrid system combines a conventional internal combustion engine and an electric motor. Both power sources work together in order to compensate for each other's' shortcomings. For electric vehicles, the combustion engine is not required. There are various types of electric and hybrid vehicles which are described in much detail in related papers, e.g. Toyota [75]. Important for the discussion here is, that the electric motor usually uses REE based permanent magnets. Primarily NdFeB magnets are used which show a relatively low Curie-temperature as described above. In order to enhance the coercivity Dy is added to replace Nd partially. Gutfleisch et al. [76] discuss suitable magnets with a proportion of around 10 % of Dy and 22 % Nd for hybrid motor use to remain stable up to about 180 °C. A Toyota Prius uses approximately 1.3 kg of NdFeB magnets arranged in a V-shape at the rotor [76]. With a share of 10 % Dy and 22 % Nd this relates to about 130 g Dy and 286 g Nd per motor. The Öko-Institute comes up with somewhat different numbers and magnet composition as is shown in Table 6.7.

The need for REE in an electric motor here ranges from about 300 g up to about 710 g of Nd.

Another important point to consider is the estimates for future developments. These data do not actually reflect absolute data but intentions. Whether they are realistic is of course another question. As will be shown later in this chapter, especially for the e-mobility, considerable incentives seem to be necessary to push the initial startup.

6.5.2 E-Mobility: Status Quo

It is widely accepted that at present the share of electric vehicles is low and that the visions are as well rather small, compared to the global annual car production. In 2009 the world production was about 62 Mio. vehicles thereof 51 Mio. passenger

Table 6.7 Comparison of given magnet data for e-motors

	Total magnet weight	Nd		Pr		Dy		Tb		Sum REE in magnet	
Öko-Institute (2011): e-motor small	998 g	**150 g**	15 %	**50 g**	5 %	**90 g**	9 %	**9 g**	1 %	299 g	**30 %**
öko-Institute (2011): e-motor big	2,373 g	**360 g**	15 %	**120 g**	5 %	**210 g**	9 %	**21 g**	1 %	711 g	**30 %**
Toyota/[75]	**1,300 g**	286 g	**22 %**	Not given		130 g	10 %	Not given		416 g	**32 %**

Notes
Bold numbers are given numbers; *regular fonts* indicate calculated values, *percent value* represents the respective ratio of the weight value

cars. In 2010 the production increased considerably to nearly 78 Mio. vehicles total, thereof 64 Mio. passenger cars [77]. As of January 1, 2012 in Germany about 43 Mio. passenger cars and in total nearly 52 Mio. cars were registered [103, 104]. Next to these production and inventory numbers, the annual sales, which usually resemble the production numbers, give an interesting insight in the todays' market penetration of hybrid technology. The figures of electric or hybrid car sales in Germany and the US from 2001 until 2011 are shown in Fig. 6.14. As a comparison, the car sales in Germany of gasoline and diesel cars have been added for a contrast.

It can easily be seen, that in the US from 2008 onwards roughly 300,000 hybrid cars were sold per year. In Germany the hybrid car sales statistics starts in 2005 with 3,589 cars sold. In 2011 just 12,622 hybrids were sold, compared to about 1.6 Mio. gasoline and 1.5 Mio. diesel cars. The development of the prestigious Toyota PRIUS started in 1995 before the global market entry in 2000 set a new milestone in mass market technology. In 2011 one million Prius cars were sold cumulative in the US and two million were sold globally [78]. In Japan alone 210,000 cars were sold in FY 2011, mainly due to governmental stimulus packages [79, p. 56].

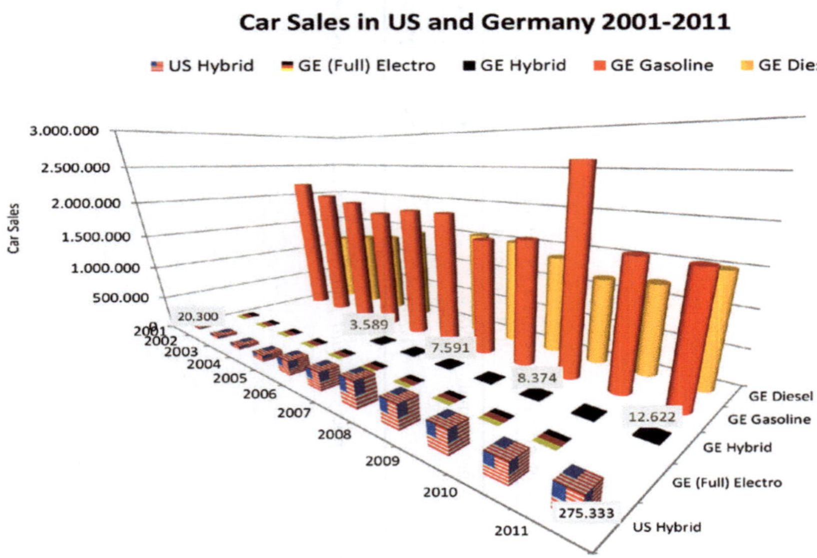

Fig. 6.14 Car sales in US and Germany 2001–2011. *Data* US 2001–2005: Duly [96]. 2006–2007: Hybrid Car Product Team [98]. 2008–2009: Brosnan [95]. 2010–2011: Hybrid Car Product [99]. Germany: KBA [102]

6.5.3 Data Analysis E-Mobility

With these numbers in mind, the visions of Germany, that until 2020 one million electric cars shall drive on German roads [80, p. 10] or President Obamas vision of 1 Mio plug-in Hybrids in five years' time (i.e. 2014) [81], are rather modest; at least considering the global annual production units and car sales. Compared to the registered hybrid or electric car numbers of today however, the visions remain challenging. It seems that the acceptance in the population for electromobility in theory is given, but to suit the action into word seems only possible, when incentives or grants are offered [82, 83, p. 7]. For this the governments of several countries started ambitious programs to support electro mobility [84]. In Germany a research group, the Nationale Plattform Elektromobilität (NPE), has been established to pursue the governmental objectives. The NPE issued same predictions about the development of electric cars in Germany. In 2011 the projection starts with 10,000 electric vehicles to reach 200,000 vehicles in inventory in 2015; 500,000 in 2017 before in 2020 one million shall be achieved [83, p. 32]. The nomenclature electric vehicle here includes both pure battery electric vehicles (BEV), Plug-in-Hybrid vehicles (PHEV) and Range extender electric vehicles (REEV). A similar prediction has been done by Becker [82] for the US as is shown in Fig. 6.15.

Attention is required when looking at the table given by the NPE [83] in their report; it shows the desired *inventory* numbers, i.e. 1 Mio. electro-cars cumulative shall be in inventory in 2020 which is not equivalent to the desired annual car sales. The actual car sales data can of course be easily derived from the given data. It shall be mentioned here, that these seemingly small difference in wording can lead to confusion or even misunderstandings.

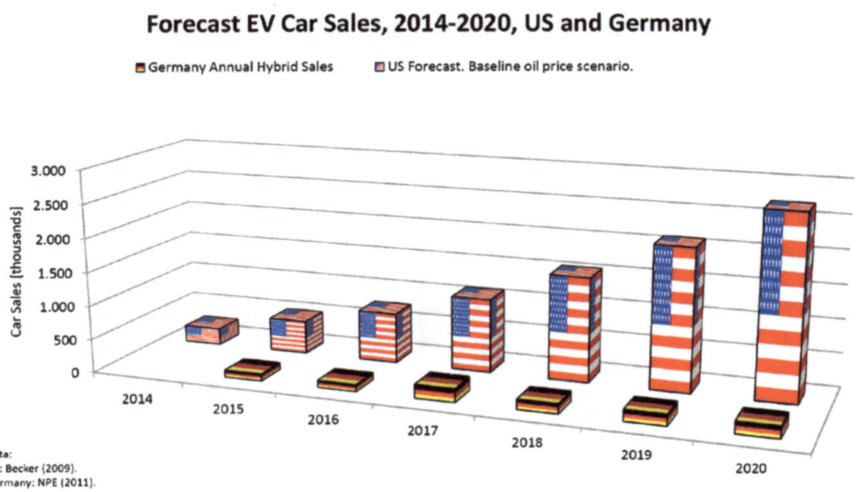

Fig. 6.15 Forecast EV car sales, 2014–2020, US and Germany

For Germany in 2020 the annual hybrid car sale would be in the order of 180,000 cars. If the annual car sales numbers in Germany remain in the order of 3 Mio. automobiles, the objective of 180,000 electric car sales in 2020 remains a modest number. Interestingly enough did a GM (General Motors) spokesman announce a temporarily halt to production of the Volt-model until April 23 [85]. The reluctance to buy electric vehicles is driven by complaints from (potential) consumers about the high prices for the cars, about some safety concerns in respect to the batteries, that the required infrastructure is not readily available and that the battery recharging lasts too long.

To get back to the REE issue, a small graph shall illustrate the required share of Nd and Dy in electric vehicles up to date. For this the car sales numbers for the US and Germany have been used as other global sales data were not attainable. So it has to be kept in mind, that the numbers shown do not represent the global share, but a part thereof. For the magnet weight a PRIUS type e-motor magnet as an overall average has been assumed.

Figure 6.16 shows the amount of Nd and Dy that was required to produce the sold cars in the US and Germany together based on the details given in the figure. The cumulative value is remarkable as within the last 10 years about 625 t of Nd and 284 t of Dy were required to produce the sold cars in Germany and the US. In 2010 alone the hybrid car sales in the US (253,719 units) and the German markets (11,202 units) accounted for about 76 t of Nd and about 34 t Dy.

Obviously here only the US and German markets were addressed. For other global markets only very partial sales and production data could be obtained. For 2010 Toyota states that nearly 600,000 hybrid cars have been produced in Japan [86]. With the same assumptions as taken above, to produce these 600,000 Hybrids it took about 190 t Nd and 86 t Dy, including the German and US sales. Still there

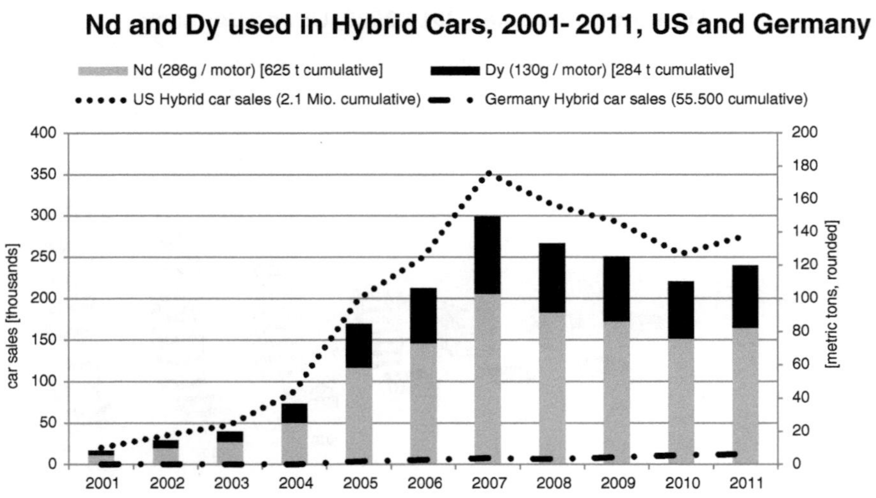

Fig. 6.16 Nd and Dy used in hybrid cars, 2001–2011, US and Germany

are some companies missing like Ford, Honda, Hyundai, Porsche, BMW and Daimler which certainly produced hybrid and electric cars in markets other than the US and Germany. For these companies no reliable data could be obtained.

As a short interim summary it can be said that in 2010 in the US and Germany together about 265,000 electric/hybrid cars have been sold. To manufacture these cars, about 76 t Nd and 34 t Dy have been required assuming 286 g Nd and 130 g Dy per car. A very rough calculation of the 2011 sales data of hybrids in the US results in a share of 65 % for Toyota hybrids (165,000 cars) which represent about 27 % of the 600,000 hybrids produced by Toyota in 2010. Adding to that share the unknown Toyotas of the German market the overall share of both markets can be set at about 30 % for simplicity reasons. Now this 30 % share is adopted to the overall hybrid sales, i.e. the 265,000 cars. Then the global hybrid sales sum up to about 850,000 cars in 2010. Under this assumption the total quantity of Nd required to produce all electric/hybrid cars in 2010 is in the order of 250 t Nd and 115 t Dy.

6.5.4 Outlook E-Mobility

For the future predictions two versions have been calculated: one being the small e-motor version and the other a big motor version using data from the Oeko-Institut [74]. Finally a share of 50 % small and big each has been determined to get an idea of the required REE amount in future for this usage.

Figure 6.17 shows the amount of Nd and Dy which will be manufactured into the predicted electric cars to be sold within the next 8–9 years. In the graph an average value has been assumed which is about one-third bigger than the PRIUS motor type today. It cannot be estimated, if the average value is reasonable or too high or too low. Despite this uncertainty, a scale ratio is obvious. In 2020 about 1,250 t Nd and about 430 t Dy are required to produce the predicted cars for the US and German market. The question is 'how big is the global market' or 'which shares do the US and German market represent'? Assuming the considerations made above that the two markets account for about 30 % of the global market, it can be assumed that the amount of Nd required in 2020 is in the order of 4,200 t Nd and 1,450 t Dy for the electric traction motors alone.

6.5.5 Further Research

This chapter showed a lot of uncertainties and had to be built on assumptions which remain a point for criticism. However, to resolve this uncertainty, it would require quite some effort and research which could be subject for further research. First detailed, accurate and reliable data is required about actual quantities of REE built into e-mobility vehicles sorted along applications like glass, small e-motors,

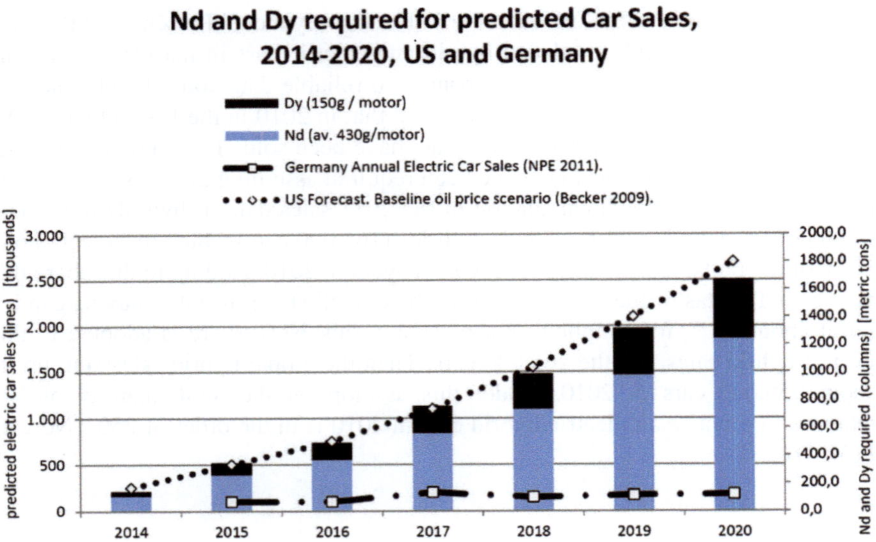

Fig. 6.17 Nd and Dy required for predicted car sales, 2014–2020, US and Germany

catalyst etc., but also in respect of passenger cars, utility vehicles and so on. This has to be evaluated along a timeline starting from about the year 2000 and reaching as far as national plans for e-mobility range into future. And there should be plans considered including big global players like China and India. As well a closer look at potential e-motors without the need for REE should be done.

E-mobility I hybrid cars		
2005*	172 t Nd	78 t Dy
2010	250 t Nd	115 t Dy
2020**	4,127 t Nd	1,440 t Dy

* Assuming US sales 200,000 as 30 % of global market
** Assuming US and German market predictions account for 30 % of global market

6.6 Synopsis of REPM Use

It has been said in the introduction, that the headlines and the common perception about REE is, that without them, the production of HDDs, mobile phones, wind turbines and electro mobility is at stake. The results of the detailed analyses above showed a somewhat different picture. Here now a synopsis of the different already discussed results about REPM use is given. As a starting point the average annual production of Nd in the years 2005 and 2010 is applied as there the 'best' or most data was available—despite the mentioned contradictory data. Then the respective shares of the applications HDD (VCM only), mobile phones, wind turbines and electro mobility are listed and finally subtracted from the total production value. As long as only one of the areas like HDDs is considered, the number of close to 2,000 t Nd required seems huge, in company with the annual production value and the other candidates quantities, the relationships get more obvious. The data is presented in Table 6.8.

Table 6.8 summarizes the already discussed quantities required for the four areas. Amongst these areas, in 2005, the VCMs for HDDs played the major role with 1,423 t Nd required. This corresponds to about 8 % of the global Nd production number. All four areas together accounted for about 2,000 t Nd or 11 % share. In 2010 the magnets for VCMs in HDDs still had the major share and all areas accounted for about 3,500 t or 15 % of the global Nd production. These numbers are lower than expected because the headlines led to the impression that these four areas cover the major share of the global production.

Of course there are several uncertainties in these figures and numbers. At least for the own measurement of HDD magnets the data base for the years from 2007 onwards did not contain enough samples for an accurate measure. And not all magnets could be analyzed for their composition in a laboratory using RFA. And not all magnets which are manufactured in HDDs and computers were included, e.g. spindle drive magnets, CD-/DVD drives and maybe cooler fans. It is not known whether all of these possible candidates actually contain REE based magnets or just some other PM. For the mobile phones the state of the art smart phones were not available for analysis. For both areas, HDD and mobile phones, the global production values had to be taken from internet-available market analyst data which could only partially be verified. In the wind industry again not all company reports did give exhaustive information about their products, productions and sales. The weight of the magnets had to be derived from scientific literature and could not be verified with the companies themselves. The same is applicable for the e-mobility issue. As the registration data for Germany and the US can be considered precise, the rest of the world data had to be extrapolated. So there remain several uncertainties.

It is also obvious that the derived figures for producing the relevant magnets do not include any production scraps and discards which are estimated to be at around 10 % [87]. These quantities would have to be added to attain the total produced elements. Nevertheless, the relations can easily be seen. And even when 20 % would be added to the figures derived in this work, there still remain considerable deltas. Table 6.8 finally shows these deltas of about 16,000 t for 2005 and about 19,000 t for 2010. This is the really interesting and also worrisome message!

Not the thought of main uses HDD, iPods, wind and e-mobility, are responsible for the major share of Nd use, instead there is a huge number or about 80–85 % of which it is not known, where the quantities go to. This number is or should be of special interest as the applications manufactured herewith would really be hit by a potential shortage. Even though a long list of magnet applications is known today and listed in Sect. 5.2, the real magnitude of these uses seems not to be in the awareness of todays' discussions. So here one of the most urgent needs for further

Table 6.8 Synopsis of Neodymium use

		Neodymium use					
		2005		2010		2020	
Supply	Global production (mean average)	17,904 t	Global share	22,256 t	Global share	n/a (48,000 t[ck])	
Demand	Global total *HDD VCM*	1,423 t	8 %	1,818 t	8 %	n/a (2,700 t[z1])	n/a (6 %)
	Global total *Mobile phones*	230 t	1.3 %	4,50 t	2 %	n/a (1,500 t[z2])	n/a (3 %)
	Global total *Wind turbines*	100 t[e]	<1 %	804 t	3.6 %	5,200 t	n/a (11 %)
	Global total *e-mobility*	172 t	1 %	2,50 t	1.1 %	4,127 t	n/a (9 %)
	Sum	1,925 t	10.8 %	3,322 t	14.9 %	n/a (13,527 t)	n/a (28 %)
Delta	(rounded to nearest 100, respectively full %)	*16,000 t*	*89 %*	*18,900 t*	*85 %*	*n/a (34,473 t)*	*n/a (72 %)*

Notes
e estimated
n/a no data available
(…) data in brackets are very rough estimates only. Percentages are based on these estimates. All bracketed data require additional research!
ck Chegwidden and Kingsnorth [94]. Value estimated for 2015 in the magnets application field
*z*1 estimate based on the assumption that in 2020 1.2 billion HDDs are produced, thereof 800 Mio. as 3.5″ with 12 g average magnet weight containing 25 % Nd, plus 350 Mio. 2.5″ HDDs with 2.5 g magnet weight containing 25 % Nd
*z*2 estimate based on the assumption that the amount of mobile phones sold increases by 76 % from 2015 to 2020 (from 2005 to 2010 an increase of 96 %, from 2010 to 2015 an increase of 86 % was noted). Figure rounded

research is hidden: to determine the quantitative shares for the individual applications in the permanent magnet arena. This is valid for other REE as well, especially Dy and Tb and to a lesser extent Pr and Sm as other important magnet ingredients. Only with these quantities in mind, possible future demands can be made transparent and feasibility for additional mining, options for substitution, miniaturization or recycling can be efficiently determined.

One more interesting fact lays hidden in these delta figures. They contain indeed substantial and potential recycling quantities. So a mobile phone recycling for retrieving the REE may not be the best option. The same is probably true for HDD recycling in the quest for REE. But they probably can be recycled if there is other REE recycling in place so that mobile phones and HDDs can be added as feedstock. These 19,000 t of Nd (in 2010) are the real big number and most interesting for a possible recycling.

Certainly another question has to be allowed, if the average production data actually are correct? Without the answers to these questions all further considerations seem vague and without solid basis. It is known, but maybe not evident, that inaccuracies exist and a lot of assumptions are being made. Even in this work. It has been tried to emphasize, highlight and analyze the discrepancies and uncertainties. Maybe a side note but it still deserves some thought, why these discrepancies exist even though most, if not all, data sources can be traced back to data published by the CREIC out of Baotou, Inner Mongolia, China?

Let's pretend that the average figures derived in this work are about correct, then finally there remains the question and the really strong quest for the search of 19,000 t of Nd, obviously or seemingly produced in 2010 and the products where they were built into. Whereas the 2010 data is of general interest the production data of the years 2000–2005 (and maybe before) are probably of way more interest as these products could presently reach their end of life date and would be available for further life or re-cycles.

References

1. Western Digital (2011) 2011 Annual report and form 10-K. Retrieved from https://materials.proxyvote.com/Approved/958102/20110916/AR_102301/images/Western_Digital-AR2011.pdf. Accessed 11 Nov 2012
2. Schüler D, Buchert M, Liu R, Dittrich S, Merz C (2011) Study on rare earths and their recycling: final report for the greens/EFA group in the European parliament, Darmstadt. Retrieved from http://www.oeko.de/oekodoc/1112/2011-003-en.pdf. Accessed 11 Nov 2012
3. Western Digital (2008) Annual report 2008. Retrieved from https://materials.proxyvote.com/Approved/958102/20080917/AR_27910/HTML1/default.htm. Accessed 11 Nov 2012
4. Gaudin S (2010) PC shipments set to grow 22 % in 2010, Gartner says—Computerworld. Retrieved from http://www.computerworld.com/s/article/9177392/PC_shipments_set_to_grow_22_in_2010_Gartner_says. Accessed 11 Nov 2012
5. Hitachi (2010) Hitachi develops recycling technologies for rare earth metals. Retrieved from http://www.hitachi.com/New/cnews/101206.pdf. Accessed 11 Nov 2012

6. Seagate (2011b) Fiscal Q4 2011. Supplemental financial information. Retrieved from http://www.seagate.com/docs/pdf/corporate/supplemental_q4_2011.pdf. Accessed 2 March 2012
7. Gartner (2005) Gartner says mobile phone sales will exceed one billion in 2009. Gartner dataquest July 2005. Press release. Retrieved from http://www.gartner.com/it/page.jsp?id=492163. Accessed 11 Nov 2012
8. Gartner (2006): Mobile phones sales up 21 % in 2005. Retrieved from http://www.informationweek.com/mobile-phone-sales-up-21-in-2005-gartner/181401126. Accessed 11 Nov 2012
9. Gartner (2008) Gartner says worldwide mobile phone sales increased 16 % in 2007. Press release. Retrieved from http://www.gartner.com/it/page.jsp?id=612207. Accessed 11 Nov 2012
10. Gartner (2009) Gartner says worldwide mobile phone sales grew 6 % 2008. Press release. Retrieved from http://www.gartner.com/it/page.jsp?id=1126812. Accessed 11 Nov 2012
11. Gartner (2011) Smartphone sales grew 72 % in 2010. Press release. Retrieved from http://www.gartner.com/it/page.jsp?id=1543014. Accessed 11 Nov 2012
12. Llamas RT, Drake SD, Stofega W, Crook SK (2011) Worldwide mobile phone 2011–2015 forecast and analysis: an IDC study. Abstract. Study available from http://www.idc.com/getdoc.jsp?containerId=232163. Accessed 11 Nov 2012
13. Heier S (2009) Windkraftanlagen: Systemauslegung, Netzintegration und Regelung; mit 15 Tabellen, 5th edn. Vieweg + Teubner, Wiesbaden
14. Jensen BB (2010) Why use permanent magnet generators in wind turbines? In: 2010 Danish society for wind energy annual conference, Bredsten. Retrieved from http://www.vindselskab.dk/composite-2334.htm. Accessed 3 April 2012
15. Polinder H, Bang D, van Rooij RP, McDonald AS, Mueller MA (2007) 10 MW wind turbine direct-drive generator design with pitch or active speed stall control. In: IEEE 2007, 1-4244-0743-05/07
16. Polinder H, Bang D-J, Li H, Chen Z (2007) Project upwind. Integrated wind turbine design: concept report on generator topologies, mechanical & electromagnetic optimization. Deliverable No.: D 1B2.b.1. Retrieved from http://www.upwind.eu/media/811/Deliverable_1B2.b.1.pdf. Accessed 11 Nov 2012
17. Shrestha G, Polinder H, Bang D-J, Ferreira JA (2010) Structural flexibility: a solution for weight reduction of large direct-drive wind-turbine generators. IEEE Trans Energy Conv, vol 25(3). doi:10.1109/TEC.201.2048713
18. Bywaters G, John V, Lynch J, Mattila P, Norton G, Stowell J (2004), (Northern Power Systems), Salata M (General Dynamics Electric Boat), Labath O (Gear Consulting Services of Cincinnati), Chertok A, Hablanian D (TIAX) (2004) Northern Power Systems WindPACT drive train alternative design study report: 12 April 2001–31 Jan 2005. Retrieved from http://www.nrel.gov/docs/fy05osti/35524.pdf. Accessed 11 Nov 2012
19. Bradsher K (2011) China consolidates grip on rare earths. Retrieved from http://www.nytimes.com/2011/09/16/business/global/china-consolidates-control-of-rare-earth-industry.html?pagewanted=all. Accessed 11 Nov 2012
20. Hollins O (2010) Lanthanide resources and alternatives. Retrieved from http://www.oakdenehollins.co.uk. Accessed 11 Nov 2012
21. Seaman J (2010) Rare earths and clean energy: analyzing china's upper hand: note de l'Ifri
22. BTM Consult (2010) World market update 2009: BTM projects 1.000 GW of wind power by 2019. Press Release. Retrieved from http://btm.dk/public/Pressrelease_Eng_BTM_WMU2009.pdf. Accessed 11 Nov 2012
23. BTM Consult (2011b) BTM consult releases new wind report: world markets update 2010: Press release 28 March 2011
24. Enercon (2011) Debatte um Neodym—ENERCON nutzt umstrittenes Material nicht. Windblatt, vol 2, p 4
25. GWEC (2011) Global wind report. Annual market update 2010. Brussels. Retrieved from www.gwec.net. Accessed 11 Nov 2012

26. WWEA (2010) World Wind Energy Report 2009, Bonn. Retrieved from http://www.wwindea.org/home/images/stories/worldwindenergyreport2009_s.pdf. Accessed 11 Nov 2012
27. Vestas (2011a) Annual report 2010. Retrieved from http://www.vestas.com/en/annual-report-2010.aspx?action=3. Accessed 11 Nov 2012)
28. Suzlon (2011) Annual report 2010–2011: sustainable pursuit of excellence
29. Gasch R, Twele J (2011) Windkraftanlagen: Grundlagen, Entwurf, Planung und Betrieb, 7th edn. Vieweg + Teubner, Wiesbaden
30. Enercon (2010) ENERCON wind energy converters. Technology & Service, Company brochure, Aurich
31. Vacuumschmelze (2010) Windprognose. Trends in der Generatoren-Entwicklung für Windenergieanlagen. Antriebstechnik 8:24–26. Retrieved from http://www.vacuumschmelze.de/fileadmin/Medienbiliothek_2010/Downloads/DM/Windenergie.pdf. Accessed 11 Nov 2012
32. Rollik D, Schleede B (2007) Nd-Fe-B magnets are going offshore in wind mill generators. Bodo's power systems, vol 11, pp 30–31. Retrieved from retrieved via http://www.vacuumschmelze.com. Accessed 11 Nov 2012
33. Windpower (2008) Getting out of gear and into magnets. Windpower monthly magazine. Retrieved from http://www.windpowermonthly.com/news/962701/Getting-gear-magnets/. Accessed 11 Nov 2012
34. Henriksen M, Jensen BB (2011) Induction generators for direct-drive wind turbines, pp 1125–1130
35. Heier S (2010) Electrical systems of WTG. Wind power grid integration in China and Europe. Wind power Asia 2010 conference, Beijing
36. Lifton J (2009) The rare earth crisis of 2009. Detroit. Retrieved from http://www.techmetalsresearch.com/2009/11/the-rare-earth-crisis-of-2009-part-2-the-green-wind-blows-from-china/. Accessed 11 Nov 2012
37. Quilter J (2011) Clipper halts 10 MW Britannia turbine. Windpower monthly magazine. Retrieved from http://www.windpowermonthly.com/channel/technology/news/1085669. Accessed 11 Nov 2012
38. Henriksen M, Jensen BB (2011) Induction Generators for Direct-Drive Wind Turbines. In: Proceedings of the IEEE International Electric Machines & Drives Conference (IEMDC), 2011
39. Hopwood D (2011a) Generation innovation for wind turbines. Renew Energy Focus 12(2): 36–41. Retrieved from http://www.renewableenergyfocus.com/view/19365/generation-innovation-for-wind-turbines/. Accessed 11 Nov 2012
40. AMSC (2010a) SeaTitan TM 10 MW wind turbine: a white paper by AMSC. November 2010. Retrieved from www.amsc.com. Accessed 11 Nov 2012
41. Hopwood D (2011) Vestas' Engel: "we believe the gearbox is the best solution…": David Hopwood speaks to Vestas' CEO Ditlef Engel. Renew Energy Focus 12(2):42–44
42. Vestas (2011c) V112-3.0 MW onshore. Online Brochure. Retrieved from http://nozebra.ipapercms.dk/Vestas/Communication/Productbrochure/V11230MW/V11230MWOnshoreDE/. Accessed 11 Nov 2012
43. BTM Consult (2011a) International wind energy development: world market update 2010: March 2011. Executive summary
44. Vestas (2011b) Turbine overview. Retrieved from http://www.vestas.com/en/wind-power-plants/procurement/turbine-overview.aspx#/vestas-univers. Accessed 11 Nov 2012
45. Vestas (2011d) V164, 7.0 MW. Online PDF, Version 03/2011-EN. Retrieved from http://nozebra.ipapercms.dk/Vestas/Communication/Productbrochure/V16470MW/V16470MW/. Accessed 11 Nov 2012
46. Vestas (2011e) Vestas' V112-3.0 MW reaches important milestone–first GW of orders already secured: Company announcement 39/2011 from Vestas Wind Sytems A/S. Retrieved from http://www.vestas.com/files//Filer/EN/Investor/Company_announcements/2011/111006_CA_UK_39.pdf. Accessed 11 Nov 2012

47. Sinovel (2011) Company overview and milestones. Retrieved from http://sinovel.com/en/companyoverview.aspx. Accessed 11 Nov 2012
48. GE (2010) 4.0-110 Offshore-Windenergieanlage. Product brochure
49. Goldwind (2011) Innovating for a brighter tomorrow. Annual report 2010
50. Gamesa (2011a) Gamesa G10X-4.5 MW. Innovating for reliability. System brochure
51. Gamesa (2011b) Annual report 2010. Retrieved from http://www.gamesacorp.com/recursos/doc/accionistas-inversores/informacion-financiera/memoria-anual/english/annual-report-2010.pdf. Accessed 11 Nov 2012
52. Gamesa (2011c) Gamesa in the offshore market. Retrieved from http://www.gamesacorp.com/en/gamesaen/strategy/gamesa-in-the-offshore-market.html. Accessed 3 April 2012
53. AMSC (2010b) American superconductor corporation AMSC & Dongfang to develop 5 MW Offshore wind turbines, Massachusetts. Retrieved from http://www.renewableenergyworld.com/rea/news/article/2010/01/amsc-to-develop-5-mw-windturbines-with-dongfang. Accessed 11 March 2012
54. Dongfang (2010) Dongfang electric direct drive wind turbine production: listed in wind turbines manufacturer in China. Retrieved from http://windturbines.china-windturbine.com/2010/10/dongfang-wind-turbine/. Accessed 11 Nov 2012
55. RePower (2011) Annual report 2010/11
56. Siemens (2009) Offshore wind power projects
57. Siemens (2011) Wind power details. Retrieved from http://www.energy.siemens.com/hq/en/power-generation/renewables/wind-power/. Accessed 11 Nov 2012
58. Junfeng L, Pengfei S, Hu G (2010) China Wind Power Outlook 2010
59. United Power (2011) History and key milestones. Retrieved from http://gdupc.com.cn/en/contents/234/695.html. Accessed 11 Nov 2012
60. Mingyang (2011) Company profile. Retrieved from http://www.mywind.com.cn/English/about/index.aspx?MenuID=050101. Accessed 11 Nov 2012
61. Aerodyn (2011) SCD-technology. Retrieved from http://www.scd-technology.com. Accessed 11 Nov 2012
62. Nordex (2011c) Designed to perform: generation gamma–Die Nordex-Effizienzklasse: Company Product Brochure
63. Nordex (2011b) N150/6000: Die wirtschaftlichste Lösung für Offshore-Projekte. Company Product Brochure. Retrieved from http://www.nordex-online.com/fileadmin/MEDIA/Produktinfos/Nordex_Offshore_N150_de.pdf. Accessed 10 March 2012. No longer available on 11 Nov 2012
64. Nordex (2011a) Geschäftsbericht 2010. Retrieved from http://www.nordex-online.com/fileadmin/MEDIA/Geschaeftsberichte/Nordex_GB2010_DE.pdf. Accessed 11 Nov 2012
65. Mitsubishi (2011) Wind turbine product information. Retrieved from http://www.mpshq.com/products/wind_turbines. Accessed 11 Nov 2012
66. Shanghai Electric (w.y.) (2001) Wind power–welcome to Shanghai electric group! Retrieved from http://www.shanghai-electric.com/en/Pages/default.aspx. Accessed 11 Nov 2012
67. Converteam (2010) Converteam powers China's first 5 MW permanent magnet direct drive offshore wind generator in china ≫ offshore wind: XEMX has recently unveiled the world's first 5 MW permanent magnet generator for a direct-drive offshore-specific wind turbine. Retrieved from http://www.offshorewind.biz/2010/11/30/converteam-powers-china%e2%80%99s-first-5-mw-permanent-magnet-direct-drive-offshore-wind-generator-in-china/. Accessed 11 Nov 2012
68. Clipper (2011a) Liberty 2.5 MW Windturbine: Company Brochure. Retrieved from http://www.clipperwind.com/pdf/Liberty_Brochure_2011_V3_Single.pdf. Accessed 11 Nov 2012
69. Clipper (2011b) Profile and company info. Retrieved from http://www.clipperwind.com/. Accessed 11 Nov 2012
70. de Vries E (2011) Technology of the sea–offshore wind turbines. Retrieved from http://www.windpowermonthly.com/search/articles/phrase/technology%20of%20the%20seas%20offshore%20wind%20turbines/. Accessed 11 Nov 2012

71. GWEC (2010) Global wind energy outlook 2010. Brussels
72. Electric-Bicycle-Guide (2012) The Chinese electric bike market. Retrieved from http://www.electric-bicycle-guide.com/chinese-electric-bike.html. Accessed 11 Nov 2012
73. FAZ (2011) Elektroräder werden immer beliebter. Frankfurter Allgemeine, 19 Dec 2011, p 15
74. Oeko-Institut (2011) Ressourceneffizienz und ressourcenpolitische Aspekte des Systems Elektromobilität: Arbeitspaket 7 des Forschungsvorhabens OPTUM: Optimierung der Umweltentlastungspotenziale von Elektrofahrzeugen. Freiburg
75. Toyota (2003) Toyota hybrid system THS II. Retrieved from www.ecworld.com/library/toyotahs2.pdf. Accessed 11 Nov 2012
76. Gutfleisch O, Willard MA, Brück E, Chen CH, Sankar SG, Ping Liu J (2011) Magnetic materials and devices for the 21st century: stronger, lighter, and more energy efficient. Adv Mater 23:821–842. doi:10.1002/adma.201002180
77. VDA (2011) Jahreszahlen Automobilproduktion. Retrieved from http://www.vda.de/de/zahlen/jahreszahlen/automobilproduktion. Accessed 11 Nov 2012
78. Toyota Motor North America (2011) Hybrids: a core technology for the future. Retrieved from http://www.toyota.com/about/environment/hybrids/index.html. Accessed 11 Nov 2012
79. Toyota Motor Corporation (2011) Annual report 2011: year ended March 31, 2011. Retrieved from http://www.toyota-global.com/investors/ir_library/annual/pdf/2011/. Accessed 11 Nov 2012
80. Bundesregierung (2011) Regierungsprogramm Elektromobilität. Berlin
81. Obama B (2008) New energy for America: transcript of speech held on 4 Aug 2008. Retrieved from http://www.pbs.org/newshour/bb/politics/july-dec08/obamaenergy_08-04.html. Accessed 11 Nov 2012
82. Becker TA (2009) Electric vehicles in the United States. A new model with forecasts to 2030, Berkeley
83. NPE (2011) Zweiter Bericht der Nationalen Plattform Elektromobilität. Berlin. Retrieved from http://www.forum-elektromobilitaet.de/assets/mime/f72061dd10e34d52e6aaaa79bdb110ab/RZ_0905_NPE_Zweiter-Bericht_Final_Print.pdf. Accessed 11 Nov 2012
84. VDA (2010) Förderung der Elektromobilität international 2006–2020. Retrieved from http://www.vda.de/de/publikationen/publikationen_downloads/detail.php?id=961. Accessed 11 Nov 2012
85. Reuters (2012) GM to halt volt production for five weeks. Detroit. Retrieved from http://www.reuters.com/article/2012/03/02/gm-volt-idUSL2E8E2CJP20120302. Accessed 11 Nov 2012
86. Toyota (2012) Number of vehicles produced in Japan by model. Retrieved from http://www.toyota-global.com/company/profile/figures/production_en.html. Accessed 11 Nov 2012
87. Du X, Graedel TE (2011) Global in-use stocks of the rare earth elements: a first estimate. Environ Sci Technol 45:4096–4101. doi:10.1021/es102836s
88. Disktrend (2000) Five decades of disk drive industry firsts. Retrieved from http://www.disktrend.com/5decades2.htm. Accessed 18 March 2012
89. McKendrick D (1997) Sustaining competitive advantage in global industries: technological change and foreign assembly in the hard disk drive industry. Report 97-06
90. Seagate (2011a) Company milestones. Retrieved from http://www.seagate.com/www/en-us/about/corporate_information/company_milestones. Accessed 11 Nov 2012
91. Seagate (2011c) Seagate and Samsung announce broad strategic alignment: Samsung to combine hard disk drive operations into seagate companies... Financial Press Release. Retrieved from http://www.seagate.com/ww/v/index.jsp?locale=en-US&name=samsung-seagate-alignment-announce-pr&vgnextoid=d00a78162ab6f210VgnVCM1000001a48090aRCRD. Accessed 11 Nov 2012
92. Toshiba (2011) 2010 Annual report. Operational review. Retrieved from http://www.toshiba.co.jp/about/ir/en/finance/ar/ar2010/tar2010e.pdf. Accessed 11 Nov 2012

93. Western Digital (2010) Annual report 2010. Retrieved from https://materials.proxyvote.com/Approved/958102/20100916/AR_67523/images/Western_Digital-AR2010.pdf. Accessed 11 Nov 2012
94. Chegwidden J, Kingsnorth DJ (2010) Rare earths: facing the uncertainties of supply: 6th international rare earths conference. Hong Kong, Nov 2010. Retrieved from http://www.slideshare.net/Tehama/rare-earths-facing-the-uncertaintys-of-supply. Accessed Nov 11 2012
95. Brosnan (2012) Reporting 2011 US hybrid car sales. Retrieved from http://hybridcar.com. Accessed 20 March 2012. Webpage ceased service Oct 18, 2012. New access via www.hybridcars.com)
96. Duffy (2008) Hybrid car sales figures. Retrieved from http://hybridcar.com/index.php?option=com_content&task=view&id=565&Itemid=45. Accessed 20 Mar 2012. Webpage ceased service 18 Oct 2012. New access via www.hybridcars.com
97. Heise (2010) Festplattenmarkt legt vor allem dank der 2.5-Zoll-Modelle zu. Retrieved from http://www.heise.de/resale/meldung/Festplattenmarkt-legt-vor-allem-dank-der-2-5-Zoll-Modelle-zu-943152.html. Accessed 11 Nov 2012
98. Hybridcar Product Team (2008) US hybrid car sales figures For 2007 continue upward trend. Retrieved from http://hybridcar.com/index.php?option=com_content&task=view&id=565&Itemid=45. Accessed 20 March 2012. Webpage ceased service 18 Oct 2012. New access via www.hybridcars.com
99. Hybrid Car Product Team (2010) Reporting US 2009 hybrid car sales. Retrieved from http://hybridcar.com/index.php?option=com_content&task=view&id=758&Itemid=45. Accessed 20 March 2012. Webpage ceased service 18 Oct 2012. New access via www.hybridcars.com
100. Isuppli (2010) HDD and ODD revenue was restored in 2010: Article in it computer zone, 08 May 2010. Retrieved from http://itcomputerzone.com/hardware/hdd-and-odd-revenue-was-restored-in-2010.html. Accessed 5 March 2012
101. Isuppli (2006) HDD sales to slow after record 2005. (Press release 12.04.2006 published in eetimes). Retrieved from http://www.eetimes.com/electronics-news/4059835/Report-HDD-sales-to-slow-after-record-2005. Accessed 11 Nov 2012
102. KBA (2012) Statistik, Emissionen, Kraftstoffe. Retrieved from http://www.kba.de/cln_030/nn_125398/DE/Statistik/Fahrzeuge/Bestand/EmissionenKraftstoffe/emissionen__node.html?__nnn=true. Accessed 11 Nov 2012
103. KBA (2012a) Der Fahrzeugbestand im Überblick am 1. Januar 2012 gegenüber 1. Januar 2011. Retrieved from http://www.kba.de/cln_033/nn_125398/DE/Statistik/Fahrzeuge/Bestand/2011__b__ueberblick__pdf,templateId=raw,property=publicationFile.pdf/2011_b_ueberblick_pdf.pdf. Accessed 11 Nov 2012
104. Seagate (2011d) Form 10-K (Annual Report): Filed 08/17/11 for the Period Ending 07/01/11
105. TrendFocus (2004) 300 Million HDDs shipped: reported in: business wire 05 Feb 2004. Retrieved from http://www.businesswire.com/news/home/20040205005864/en/300-Million-HDDs-Ship-2004-TrendFOCUS-Study. Accessed 11 Nov 2012
106. Western Digital (2007) Annual report 2007. Retrieved from http://www.wdc.com/wdproducts/library/company/annualreport/2278-001005-A03.pdf. Accessed 11 Nov 2012
107. Western Digital (2000) Annual report 2000. Retrieved from http://www.wdc.com/wdproducts/library/company/annualreport/annual00.pdf
108. Western Digital (2001) Annual report 2001. Retrieved from http://www.wdc.com/wdproducts/library/company/annualreport/annual01.pdf. Accessed 11 Nov 2012
109. Western Digital (2002) Annual report 2002. Retrieved from http://www.wdc.com/wdproducts/library/company/annualreport/annual02.pdf. Accessed 11 Nov 2012
110. Wong P-K (1999) The dynamics of HDD industry development in Singapore: University of California, San Diego, Report 99-03

Chapter 7
The Geography of the REE

As a last step in this work the initial picture of a *Geography of the REE* shall be drawn. The initial idea was to get a world map with the locations of REE starting from the geological side, i.e. the physical occurrences of the minerals and then proceeding via the production steps to manufacturing of the consumer products to their use. Eventually the REE get back into the production cycle or end up dissipating or getting stored somewhere. This map would of course get a good idea about the recycling potentials, both in space and time. However the analysis of this work could only discover a modest share of Nd use so that such a map in this state of research can only be very approximate and incomplete. Despite these shortcomings such a map has been tried to set up; kind of as a starting point with the intent to amend the data in the succession of further research.

The graph (Fig. 7.1) shows a world map with the population densities in the center. The populated areas are these where potentially REE are widespread used. So they offer the most promising areas for sales, recycling, dissipation, storage and also research. On the outer border a schematic life cycle chain of the REE is shown starting lower left with mining (neglecting exploration etc.), followed by the complete cycle until the REE are finally dissipated, turned back into another cycle or get stored somewhere, under controlled or uncontrolled conditions. As well the thought may arise that they get lost during recycling when other metals are primary recycling objects.

On the lower right side there are red squares indicating 1,000 t each of (probably) mined Nd (in oxidic form). The orange colored squares indicate the measured and researched shares that were found in this work. It is very obvious that these four areas only comprise a modest share of all the Nd produced. So the informational dilemma gets clear. Either the production data are very false or there actually is a huge grey number of applications and uses which are not obvious. They are of utmost importance and ask for more transparency and thus further research.

The squared letters indicate where the products are dispersed over the globe. This only gives a rough qualitative and very rough quanitative picture. It would be of interest to detail the areas further down to more restricted geographical areas.

V. Zepf, *Rare Earth Elements*, Springer Theses,
DOI: 10.1007/978-3-642-35458-8_7,

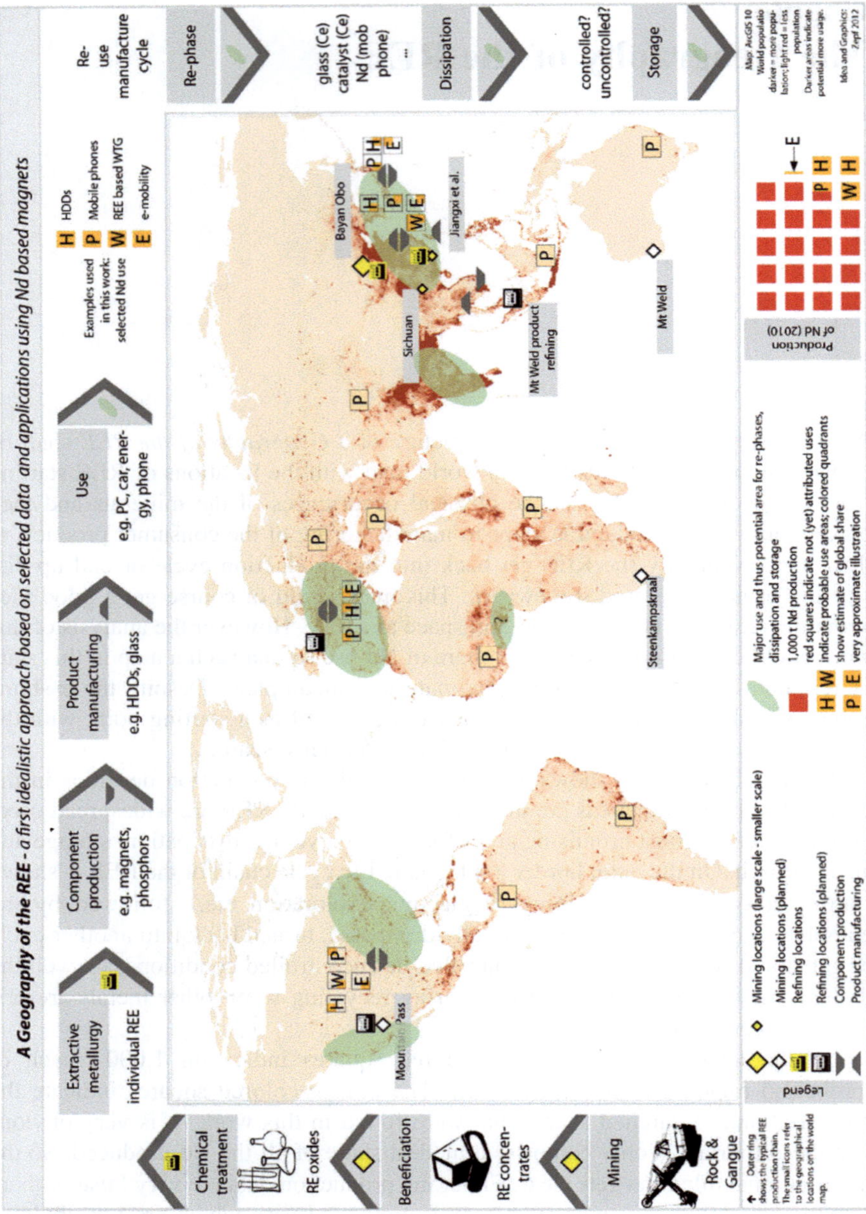

Fig. 7.1 A geography of the REE—a first approach

Thus potential quantities for recycling could be determined. Of course also other geographical issues could then be revealed, e.g. if there is the required infrastructure in place to collect the products up to the determination of necessary incentives for the people to return their old and used products for Re-phasing.

Today there is an obvious agglomeration of all life cycle areas in China and the Far East. Especially the REPM based WTGs are concentrated in China, not in the West (yet). Of course the red squares are of interest and the picture will most probably change within the next few years, so that this geography requires a progressive update both on the macro but as well on the micro scale.

Chapter 8
Conclusion: The REE and the Real Problem

The United States declare trade war against China was a headline dated March 13, 2012 on the German internet news platform N-TV [1] which broached the issue of a newsworthy claim of the US, the EU and Japan against China concerning their restrictive handling of trade with REE, molybdenum and tungsten. This approach was expected as a succession of the WTO claims described in the chapter about China above. It was not as surprising as the headlines might imply. Nevertheless the reporting remained on a top level with known arguments. So there seems to be a real problem concerning the REE. It has also been discussed above that the static range of the REE is in excess of 850 years and subject to a further increase of range as lots of exploration teams show promising results. So this unfolds that there is no physical or geological issue but something else.

Still there are mindsets speaking of strategic and unfair plans of the Chinese to lure the west into the trap: China produced REE so cheap that the West had to shut down all its ventures until after closure of the last mine, China increased prices and used its monopolistic position to rule the rest of the world; this or similar remarks have been heard during several discussions which have not been transcribed and are as such not scientifically reliable.

China lowered the export quotas which are in place since at least the year 2000. Associated with the continued reduction of quotas, the prices for REE exploded, even though the actual demand did not explode. This can be extracted from the figures elaborated in this work. As such the price volatilities must have had another reason as demand.

It also seems clear that China indeed faces extreme environmental problems in conjunction with REE mining but primarily with REE processing. And that processing is not at all trivial is widely known. Mainly the illegal mining seems to be tied to very severe environmental damaging procedures; especially in the areas of HREE mining in Southeast China. But also from Baotou considerable problems are reported by Hilsum [2] and pictures from Google earth show, what Hilsum tries to tell. So it should be in *everybody's* interest to solve the environmental problems as soon as possible. This is what China is doing, or what it pretends to do—as others would call it. These reports of the 'dirty secrets of REE mining' [4] got

V. Zepf, *Rare Earth Elements*, Springer Theses,
DOI: 10.1007/978-3-642-35458-8_8,

considerable media attention and the accusations persist. The problems addressed are however not per se associated with the REEs, instead they belong primarily (probably—another informational dilemma) to the illegal and unprofessional mining and processing associated in this case with the REE. So the headline and perception seems wrong that the REE are dirty elements. Under modern technologies the processing might be just as good or bad as any other technique but the possibilities that the REE offer are immense. So this consideration requires some more in depth analysis instead of accusations.

So there is a manifold paradox situation. There is or seems to be a shortage situation even though the static range is one of the widest known; even though the REE are found on the last third of abundance on the earth. Maybe this is a tribute that REE have not been mined for a very long time yet.

China does not have the best reputation in transparency. As well did China not fulfill all announcements as e.g. the WTO disputes described in Chap. 3 have shown. Yet, in this rather case the situation seems clear: China has a monopolistic situation, but it also has severe environmental problems and tries to solve these by means of production quotas to cut illegal mining and enforce implementation of environmental standards. In how far these intentions got realized is another question and the difficulty should become apparent when considering the geography of the huge country, the available infrastructure and political strength to enforce the politics and regulations made. As long as illegal mining finds customers, the problem however could remain a critical one. Interesting as well is that even though production quotas have been issued, there are way higher production quantities reported. It is also most obvious that China, by using the same wave, tries to get more out of the value added chain, i.e. by restrictions of export of pre-production states of REE. To say it cautiously: this seems to be a general objective in market economies.

Looking back in time, the closure of Mountain Pass and the abandoning of most of the monazite placer mining ventures mark a decisive point in time. It cannot be determined what the exact causes for the closures were, even though some hints were given in earlier chapters. For China it seems clear that they followed a long term strategy using and developing RE applications. In the West the still limited application fields of REE in the late 1990s may have contributed to the eventual closure of RE mines. In the late 1990s the green technologies and their global advance were not decisively foreseeable. Consequently China evolved as the dominant and potent RE manufacturer worldwide and a monopolistic situation developed. What makes the situation difficult is that when the mining activities were shut down not only the mines were lost, but in the course of the action also know how and specialized personnel. Today it is therefore not only a case of opening up a new mine but rather that the processing chain has to be re-invented and knowledge has to be raised—on the job. The re-invention of practical large scale REE processing, separating, refining etc. seems to be one of the most difficult and challenging task today for non-Chinese ventures. This task includes the handling of the waste chemicals from processing and the storage of potential radioactive tailings. It seems that this challenge is being faced as several promising mining projects are on the brink of production. These companies not only concentrate on the mining side but also try to offer the entire chain from mining to magnet or at least from mining to the REE. So the shortage problem, be it actual or only mental, will most probably be solved in near future. And several direct engagements of consumer companies in mining projects can be attested—as one possibility to reduce the dependency on China.

One lesson can be drawn from this experience: that the complete abandoning of a technology in favor of better prices elsewhere could prove as a big mistake in the mid to long run. The cheap prices for REE now show their real price: dependencies, environmental problems and uncertainties which drive companies and politics to search for alternatives even though the characteristics of REE show extremely positive applications, not only for daily goods, but also for green technologies.

But there still remains a complete different issue or problem: the one of the actual demand and supply of the REE. The data discussed above showed that there are a lot of questionable data sets around. As no really correct data set could be determined a lot of average values had to be derived, even though this is again vague—to build a valid (average) value based on uncertain data! So this is the biggest challenge to get transparent, reliable and valid data. This work has shown one way to get to (more) reliable data—a bottom up approach triggered by the prayer wheel-type repetitions of headlines which stress the huge problem of the REEs. The approach is quite simple—to disassemble some samples of products which contain REE and have them analyzed. Of course only a very small amount of products and samples could be processed, but the method could and should be enhanced to further decisive goods. Because it could be shown, that the initial expectation, that the areas HDD, mobile phones (or *iPods* or *iPhones*—words that the media often use), Wind and e-mobility would carry the major share of Nd use, had to be discarded. The measurements made could only determine a share of about 15 % of Neodymium on the probable global supply. Thereof HDDs hold the biggest share with about 8 % of the globally produced Nd in 2010, mobile phone magnets account for about 2 %, wind turbines for about 4 % and electromotors for e-mobility for about 1 %. Or to put it in other words using an 80/20 rule: it was thought that these areas would make up the typical 80 % whereas the rest shares the remaining 20 %. It turned out however, that the relationship is vice versa. Even when errors are assumed and the derived figures would be doubled, i.e. the 4 areas account for 30 % of the Nd, still about 70 % would be *missing*. On one side this finding is appeasing, that we do not have to fear any HDD or mobile phone shortages due to raw material issues. But on the other hand this fact shows several flaws, concerns and possibilities at the same time. The first is that there is a considerable grey number of applications which take the major share of Nd. It has to be assumed that the demand for these applications will increase with the general expected increase of the global population and thus general increase of demand. On top of this big number the described uses will have to be added. Future demand therefore could be way higher than expected today; or be lower. If e.g. the wind energy in 2020 requires 5,200 t Nd and the e-mobility about 4,200 t Nd (for the electro traction motors), as has been described above, this would be about 8,400 t more than today. Assuming an AAGR of 4.5 % (as for 2005 until 2010) from 2010 until 2020, the demand for Nd would reach about 35,000 t. On top of this the additional demand for Nd for wind turbines and e-motors would account to about 43,000 t so that the share of wind and pure e-motors would be about 24 %. This surely is a very rough calculation but it shows the scale in which things might happen. So the big players are not wind and the e-motors (alone), but rather other applications.

Without question, these findings call for further research on a multitude of levels, both on micro-, meso- and macro-levels. The quest should be to search for the missing quantities. The listings of permanent magnet applications in Sect. 5.2 give an idea of the magnitude and vastness of the research field. The idea of trying to verify at least some very probable major application areas for REPM in this chapter has been abandoned as no hot candidate could be easily identified. There are some probable candidates like the dozens of small electromotors in cars (like headlight range adjustment motors, window openers and so on) or speakers in car and home stereos. It could also be modern tools or air conditioning systems. A quick look into some of these applications shows both magnet types are present, ferrites and REE based ones, so that this research would require a completely new onset. In Germany a new project under the lead of Siemens is on the way to electro-motor recycling to retrieve mainly the REE [5]. This could be a means of getting access to the required information.

8.1 Outlook and Further Research

The widespread reported global forecast numbers usually are based on private discussions with stakeholders and specialists [3]. This is of course not at all transparent. Maybe it is not intended to reveal the actual demand in order to do politics. From a scientific point of view it gets critical when even political decisions are based on such uncertain data bases and some doubtful, incomplete knowledge.

Several options and questions for further research have been addressed throughout this work. The most important ones are the clarification of actual production data both cumulative and with special interest for the individual REE. This includes the clarification of data discrepancies. It is also important to bring some transparency in the issue, e.g. by answering questions like the need of Dy for WTG, which would instantaneously pose a huge problem as Dy is absolutely tight on supply. And finally the quantification of the uses in the application lists would bring some light into the topic and reveal the really critical applications.

The continuation of this basic research is deemed absolutely necessary to help balancing facts and opinions and thus to show the real relationships as has been initiated with this work. With the thereof derived reliable and valid results it is useful to evaluate the best suitable steps to overcome the actually persisting and potential future problems, be it more mining projects, or optimizations in processing, manufacturing, miniaturizations, design to re-phasing; extensions of the products lifespans or *re-phases* like re-use, re-manufacture or re-cycling and maybe even substitutions. All these decisions require a solid factual data base which is not seen available in a scientific sense, i.e. reliable and valid manner. There is some overarching work ongoing e.g. by Du and Graedel or the Oeko-Institute, but still additional research is absolutely necessary, not so much in the specialized areas, as there a lot of interesting research is going on already, but more on the inter—and transdisciplinary level to contribute to more transparency and to show true and real relationships and scales.

This work showed one way, how to approach such a task, with quite some interesting and not expected results.

References

1. Badtke T (2012) WTO-Streit um Seltene Erden: USA erklären China Handelskrieg. http://www.n-tv.de/wirtschaft/USA-erklaeren-China-Handelskrieg-article5743121.html. Accessed 11 Nov 2012
2. Hilsum L (2009) Chinese pay toxic price for a green world. http://www.timesonline.co.uk/tol/news/world/asia/article6946038.ece. Accessed 11 Nov 2012
3. Kingsnorth DJ, IMCOA (2010) Rare earths: facing new challenges in the New Decade, 2010. http://www.terramagnetica.com/downloads/IMCOA-2010-03-SME-Presentation-Final-R2.pdf. Accessed 11 Nov 2012
4. Shiraishi PA (2010) The rare earths revolution: The dirty secret of green technologies. http://www.cepe.ethz.ch/education/termpapers/Shiraishi.pdf. Accessed 11 Nov 2012
5. Siemens (2012) Elektromotoren recyceln: Neue Quelle für Rohstoffe—Siemens Global Website. http://www.siemens.com/innovation/de/news/2011/inno_1137_2.htm. Accessed 11 Nov 2012

Annex A
Details to the Literature Research

Study/report title Focus of study/report	Data source for production data	Data sources for share of individual REE/application	Remark Source of study
USGS mineral commodity summary			
'Statistics and information on the worldwide supply of, demand for, and flow of minerals and materials essential to the U.S. economy, the national security, and protection of the environment'	No data sources explicitly stated		http://minerals.usgs.gov/minerals/pubs/commodity/
USGS 2009 Minerals yearbook. Rare Earths [advanced release]			
As for mineral commodity summary	No explicit but general data sources: Company information • Alkane, Arafura, Grace, Jervois mining Ltd., Rare element resources Ltd. Consultants/research • BCC research • Kingsnorth, Dudley • Roskill information services Ltd. Others • China rare Earth information • Newspapers		China rare Earth information centre (CREIC) as monthly newsletter http://minerals.usgs.gov/minerals/pubs/commodity/rare_earths/myb1-2009-raree.pdf
Sinton C. W. (2006) Study of the rare Earth resources and markets for the Mt. Weld complex. For Lynas corporation Ltd. A BCC research study			
Market research for Lynas corporation Ltd	2003–2005: Morgan Stanley Neomax Report Feb. 2005, AMR technologies June 2005 prospectus, CREI Apr. 2005 and Mar. 2006; BCC research	No references given	Generally BCC research stated as reference (Sinton 2006)

(continued)

V. Zepf, *Rare Earth Elements*, Springer Theses,
DOI: 10.1007/978-3-642-35458-8,

(continued)

Study/report title Focus of study/report	Data source for production data	Data sources for share of individual REE/application	Remark Source of study
BCC Research, 2009 Rare Earths: worldwide markets, applications, technologies (market research report AVM018F), June, Wellesley, MA, BCC research, 253 p plus appendices			
• Provide an updated review of RE technology, focusing on materials, sources, and applications • Highlight new technological developments, during the past 3 years, related to RE • Review existing fields of application for RE and investigate emerging applications • Supply an updated worldwide overview of resources and RE production • Describe the market evolution since 2006 and estimate 2009 global markets by application and REE, with growth forecasts through 2014 for each market segment • Identify important technological and market trends within each market segment • Provide an updated review of current producers of RE materials and other relevant industry players • Determine trends in global patents	According to freely available information: Market analysis by analyzing 67 suppliers of REE (security and exchange commission filings, web sites, annual reports, industry directories, industry magazines, catalogues, government sources, other public sources Secondary data including USGS, patent offices, company web sites, company annual reports, stock exchange information, American Ceramic Society Publications		Cost: US $4850,—(as per homepage) Primary and secondary methodologies with different sources respective to the chapters topics (BCC Research 2009)
Liedtke M, Elsner H. (2009) Commodity top news Nr. 31. Seltene Erden. BGR Hannover, 6p			
Short paper about REE	Kingsnorth D. J. (2007) Rare Earths: an industry at the crossroads Presentation Nitschke, 2009		(Liedtke and Elsner 2009)
Nitschke, S. (2009): Brief aus bonn. Forum metall trends. 2/2009			
Short information, papers, reports about geoscientific topics. Metallurgy, secondary-recycling	USGS		(Nitschke 2009)

(continued)

(continued)

Study/report title Focus of study/report	Data source for production data	Data sources for share of individual REE/application	Remark Source of study
SATW (2010). Seltene metalle. Rohstoffe für Zukunftstechnologien. SATW Schrift 41			
Short study about rare metals shown on five examples: lithium, rare earth elements, indium, platin-group-metals, tantalum	USGS Liedtke and Elsner 2009	n/a	Swiss Academy of Engineering Sciences. Studies and presentations. Aimed upon strengthening technical innovation and knowledge for the public. Politically neutral and non-commercial oriented (SATW 2010)
Lynas Corporation Ltd. (2011). Rare Earths. We touch them everyday. Investor Presentation. May 2011			
	Industry resources and Lynas research	Non-chinese market: aggregate of estimated manufacturer demand China market: IMCOA and China rare Earth information centre	Lynas Corp. Ltd. Is an Australian company supposed to start mining and refining of rare earth products from their Mt. Weld deposit and the Lynas Advanced Materials Plant in Malaysia in late 2011, early 2012 (Lynas 2011a)
Roskill (2007) The economics of rare Earths and Yttrium. 13th ed			
• Independent research and analysis from industry experts • Market intelligence for business planning • Detailed survey of production in 25 countries • Up-to-date profiles of major production companies and their activities, including Baogang Rare Earth, Juangxi South, Rare Earth Hi Tech, Liyang Rhodia and Chevron Mining • Forecasts for end-use consumption and world supply and demand	USGS CREIC NDRC Roskill estimates	USGS CREIC Roskill	US$ 7500,—(14th edition -most actual edition) (Roskill 2007)
IMCOA – Industrial Minerals Company of Australia			
Mentioned together with Mr. Kingsnorth			No dedicated homepage known

(continued)

(continued)

Study/report title Focus of study/report	Data source for production data	Data sources for share of individual REE/application	Remark Source of study
Kingsnorth, D. (2010) Rare Earths: facing new challenges in the new decade			
Presentation about rare Earth elements written by Dudley J. Kingsnorth, industrial minerals company of Australia Pty Ltd, presented by Clint Cox, SME annual meeting 2010	Roskill 13th Ed. 'The economics of rare earths & Yttrium', November 2007 IMCOA, China rare Earth information centre (CREIC), rare Earths industry stakeholders	Grouping in application/ functionalities; no dedicated individual array: Roskill, IMCOA Demand 2008 and 2014*: IMCOA, Roskill, private discussions with industry stakeholders	'* Data accuracy +/− 15% (Kingsnorth and IMCOA 2010)
Molycorp (2011). Rare Earth resurgence: Molycorp's plan to increase global diversity in rare Earths through technology innovation			
Molycorp progress report	IMCOA and Molycorp estimates, accuracy +/−20 % (until 2015), +/−25 % (until 2020)	IMCOA	October 2011 Update. Presentation. pp. 7–8 (Molycorp and Inc 2011)
Shiraishi, P. A. (2010). The rare earth revolution: the dirty secret of green technologies. ETH Zürich, energy economics and policy			
Short paper about rare earths with unclear statement and intentions	n/a	Sinton 2006	(Shiraishi 2010)
Angerer et al. (2009) Rohstoffe für Zukunftstechnologien			
Study with focus on future technologies and raw materials needed for them	USGS, BCC research 2006	n/a	(Angerer et al. 2009)
Oakdene Hollins (2010). Lanthanide resources and alternatives			
A report for department for transport and department for business, innovation and skills, UK	Data with no year (p. 1): USGS, mineral commodity summary 2003–2009 (p. 17): USGS 2007 rare Earth yearbook 2014 data (p. 20): IMCOA	2008 and 2014 data (p. 1): IMCOA	(Oakdene Hollins 2010)
Jack Lifton (2010) The supply issue for all metals. Volume 2 Issue 4			
Basic information about metals in general	USGS	n/a	As well: Lifton, J (2010) Rare metals in the age of technology. Volume 2, Issue 1 (Lifton 2010b)
Seaman, J. (2010). Rare Earths and clean energy: analyzing China's upper hand. Note de l'Ifri			
Institute français des relations internationals (Ifri), a research center and forum for debate on major international political and economic issues	2009 data (p. 11) USGS mineral commodity summary 2010	2010 and 2014 data (p. 9) Lynas Corp. Ltd. 2010 presentation	Ifri—Governance européenne et géopolitique de l'énergie (Seaman 2010)

(continued)

(continued)

Study/report title Focus of study/report	Data source for production data	Data sources for share of individual REE/application	Remark Source of study
Byron (2010) Rare earth elements—pick your spots carefully			
Report with some basic information about rare earths characteristics and critical reflection of often stated numbers and figures	Byron capital markets	Byron capital markets	No further details to data source (Byron 2010)
USGS (2002) Rare Earth elements—critical resources for high technology			
Short paper with basic information about REE	USGS	n/a	USGS Fact Sheet 087–02, 2002 (USGS 2002b)
Hilpert H. G., Kröger A. E. (2011) Seltene Erden–die vitamine der industrie			
Compendium about raw materials and inherent conflict potentials	Generally: Liedtke/Elsner (2009) USGS Naumov	Vulcan T. (2008)	In: Mildner S.-A. [Ed.]. Konfliktrisiko Rohstoffe?. Herausforderungen und Chancen im Umgang mit knappen Ressourcen. SWP Studie. Februar 2011, Berlin (Hilpert and Kröger 2011)
Vulcan T. (2008). Rare earth metals: not so rare, but still valuable			
	USGS, Roskill	Roskill 2007	(Vulcan 2008)
CREI—China rare earth information (centre)			
Monthly newsletter with information from the CREI located in Baotou			Via subscription only
Asian Metal Ltd. (2008) Annual report on rare earth market			
Shot report about RE production and trade	None given In the text the Chinese government and the Chinese ministry of land and resources are mentioned but without a tie to the tabled numbers	n/a	(Asian Metal Ltd. 2009)
Heinritzi J. (2008) Rohstoffe. Seltene Substanzen vor Nachfrageboom			
Short investor information for rare earth investment options	108.000t produced in 2006 given without mentioning the data source		(Heinritzi 2008)
BGS (2011). World mineral production. 2005–2009. British geological survey			
World mineral production statistics	Hint to USGS and country and company homepages	n/a	(British Geological Survey 2011)

(continued)

(continued)

Study/report title Focus of study/report	Data source for production data	Data sources for share of individual REE/application	Remark Source of study
Weber L., Zsak G., Reichl C., Schatz M. (2011) World Mining Data. Volume 26			
Minerals production data	USGS, BGS, other sources	n/a	(Weber et al. 2011)
Lusty P. (2010) Critical Metals and rare earth elements			
Mines and money 2010. Presentation	2009 data (p. 8): USGS 2010, BGS	2008 data (+−15%): IMCOA 2010	(Lusty 2010)
Tse P.-K. (2011) China's rare-Earth industry. USGS Open-File Report 2011–1042			
Short but comprehensive report about China's rare Earth industry up to 2010	USGS 2011 Updated from Haxel et al. (2002)	n/a	China ministry of industry and information technology—production quota numbers China ministry of land and resources—production quota numbers Data from China ministry of commerce deals solely with export quota All Chinese sources are only available in Chinese language (Tse 2011)
Schüler et al. (2011) Study on rare Earths and their recycling			
Comprehensive Öko-Institute study	2008 and 2009 data (p. 18): USGS 2010	Share of individual REO in 2006 (p. 26ff): MEP 2009* '* Data source is not obtainable. Schüler et al. state that MEP gives following data sources: journal of rare earth information, Chinese society of rare Earths and NDRC Functional grouping: data for 2006 to 2008 (p. 62): Sources given in text: Kingsnorth 2010 (i.e., SME presentation) Great Western minerals group 2010 BGR 2009 (i.e., Liedtke and Elsner 2009)	(Schüler et al. 2011)

(continued)

(continued)

Study/report title Focus of study/report	Data source for production data	Data sources for share of individual REE/application	Remark Source of study
Reuters (2010) Fight for rare Earth			
	2009 data (p. 17): No source stated. In text BGS is mentioned	n/a	(Reuters 2010)
GWMG (2010) REE supply and demand			
	2005–2015: Kingsnorth/IMCOA 2011 In text furthermore: Roskill, IMCOA, Lynas corporation without year	n/a	(GWMG 2011)
Hurst (2009) China's rare Earth elements industry: what can the West learn?			
	2000–2012 Kingsnorth ('Dudley Chart'), taken from a presentation by Molycorp. (p. 27)	n/a	(Hurst 2010)

Mr Kingsnorth is the author of the Roskills 12 and 13th editions of the reports on rare earths (bby 2010)

Annex B
Mobile Phone Measurement Details

#	IC	Manufacturer	Type	Model	Year	T	(g)	R	(g)	V	(g)	Total (g)		Remarks
1	H9	Motorola	flap	A780	2004	2	0.4			1	0.39i	0.6	[a]	
2	H7	NEC	flap	MP6J1E1-1B	2002	2	0.3					0.3		
3	H5	Nokia	flap	RH97	2005	2	0.72i					0.72	i	
4	H6	Samsung	flap	SGHC260	2007	2	0.23			1		0.43	[a]	
5	H39	Samsung	flap	SGHX800	2005	2	0.73i	1	0.29	1	0.31	1.31	ix	
6	H14	Sharp	flap	GX30	2004	2	0.34			1	0.13i	0.47	i	
7	H8	Siemens	flap	CF62	2004	1	0.73					0.73		
8	H24	Nokia	slider	6280	2005	1	0.4i	1	0.07	1	0.06	0.53	i	1.32i all
9	H38	Nokia	slider	7110	1999			1	0.3	1		0.5	[a]	
10	H25	Samsung	slider	SGH E800	2004	1	0.76i			1		0.9	i[a]	
11	H42	Samsung	slider	SGH D900i	2006	1	0.19	1	0.31	1		0.7	[a]	
12	H60	Samsung	slider	SGH Z400	2006	1	0.15i	1	0.3			0.45	i	Was IC H43
13	H57	Sony Ericsson	slider	W580i	2007	1	0.35i	1	0.41	1		0.96	i[a]	Was IC H40
14	H49	Samsung	SmartPh	GTS5230	2009	1	0.33			1	0.04	0.37		
15	H12	Alcatel BE1	bar	16DCKJ1M		1	0.72					0.72		
16	H35	LG	bar	N98		1	0.31	1	0.29	1	0.05	0.65		Vib dose
17	H26	Motorola	bar	TCM		1	0.33			1	0.86i	0.53	[a]	
18	H30	Nokia	bar	6230i	2005	1	0.74i			1	0.23	0.97	i	1V: 1.15i
19	H34	Nokia	bar	3310	2000			1	0.34	1	1.2i	0.54	[a]	
20	H44	Nokia	bar	3310	2000			1	0.31	1	1.1i	0.5	[a]	
21	H47	Nokia	bar	5110	1998			1	0.35			0.35		
22	H43	Sagem	bar	MC930	2000	1				1		0.92		1T: Shinwoo SAS015 8Ω2
23	H40	Samsung	bar	SGH C140	2007	1		1		1		0.72		
24	H13	Siemens	bar	C35	2000	1	0.18			1	1.3i	0.38	[a]	

(continued)

V. Zepf, *Rare Earth Elements*, Springer Theses,
DOI: 10.1007/978-3-642-35458-8,

(continued)

#	IC	Manufacturer	Type	Model	Year	T	(g)	R	(g)	V	(g)	Total (g)		Remarks
25	H22	Siemens	bar	M35i	2000	1	0.16			1	0.14	0.3		Vib total 1.47
26	H23	Siemens	bar	M35i	2000	1	0.18			1	0.27i	0.45		Vib total 1.57
27	H29	Siemens	bar	S65	2004	1	0.22i			1	1.07i	0.42	i[a]	
28	H31	Siemens	bar	AX72	2005	1	0.20i			1	0.91i	0.4	i[a]	
29	H41	Siemens	bar	M55	2003	1	0.19			1		0.39	[a]	
30	H46	Siemens	bar	C35	2000	1	0.3i			1	1.22i	0.5	i[a]	
31	H48	Siemens	bar	M65	2004	1	0.21i			1	1.28i	0.41	i[a]	
32	H55	Siemens	bar	ME45	2001	1	0.2			1	0.04	0.24	x	Was IC H38
33	H59	Siemens	bar	S35i	2001	1	0.27			1	0.17	0.44		Was IC H42
34	H61	Siemens	bar	M55	2003	1	0.18			1	0.28	0.46		Was IC H44
35	H62	Siemens	bar	S35i	2001	1	0.29i			1	1.22i	0.49	i[a]	Was IC H45
36	H64	Siemens	bar	C36i		1	0.3i			1		0.5	i[a]	
37	H28	Skype	bar	WP-S1		1	0.5i			1	0.86i	0.7	i[a]	
38	H21	Sony	bar	CMDJ70	2001			1	0.28			0.28		
39	H27	Sony Ericsson	bar	K750i	2005	1	0.49	1	0.09	1	1.19i	0.78	[a]	
40	H32	Sony Ericsson	bar	T610	2003	1	0.34i			1		0.54	[a]	
41	H33	Sony Ericsson	bar	K770i	2007	1	0.39i	1	T mini	1		0.59	i[a]	
42	H36	Sony Ericsson	bar	J110i	2007	1	0.46			1		0.66	[a]	1T 0.6i
43	H45	Sony Ericsson	bar	T100	2002	1	0.24			1		0.44	[a]	
44	H56	Sony Ericsson	bar	K7701	2007	1	0.2	1	0.18i	1		0.58	i[a]	Was IC H39
											Sum	**23.27**		
											Av	**0.54**		rounded
1	H3	AKG	in ear	K 315		2	0.7					0.7		
2	H1	Ciao Roma	in ear	blau		2								
3	H16	Elecom	in ear			2	0.18					0.18	x	
4	H2	Hama	in ear	zitrone		2								

(continued)

(continued)

#	IC	Manufacturer	Type	Model	Year	T	(g)	R	(g)	V	(g)	Total (g)		Remarks
5	H4	Nokia	in ear	HS 23				2	0.43			0.43		
6	H15	Nokia	in ear	HS 23				2	0.44			0.44		
7	H18	NoName	in ear			2	0.9i					0.9	i	
8	H50	NoName	in ear					2	2.47i					Ferrit?
9	H53	NoName	in ear			2	0.67i					0.67	i	
10	H54	NoName	in ear					2	0.36i			0.36	i	
11	H20	Philips	in ear			2	0.2i					0.2	i	
12	H51	Philips	in ear	SHE3621		2	0.81i					0.81	i	
13	H17	Sony Ericsson	in ear			2	0.18							
14	H52	ThaiAir	in ear					2	2.2i					Ferrit?
15	H19	Vivanco	in ear					2	2.24i			2.24	ix	
											Sum	**4.51**		
											Av	**0.56**		rounded
	K1	AutoRes	headset			2	4.42						x	Ferrit
	K2	NoName	in ear			1	0.35						x	Ferrit?
	K3	Planitronics	headset			2	4.6i					4,6	ix	
	K4	NoName	headset			2	4.44						x	Ferrit
	K5	AKG	headset	K414P HiEnd		2	4.9					4.9	x	
	K6	Philips	headset	SBC3328		2	6.0i					6	ix	Ferrit
	H37	Casio	Camera	EXZ500		1	0.12i	1	0.07			0.19	ix	

Notes *T*—number of magnets—tablet type, *R*—number of magnets—ring type, *V*—number of magnets—vibrating motor, *i*—weight including additional materials, e.g. washer, [a] —assumed weight for vibrating motor: 0.2 g, x—excluded from calculation, *m*—missing, *IC*—internal code

Annex C
HDD Measurement Details

Label	Manufacturer	Type			Capacity		Weight of magnets (sum)	Number of VCA magnets	Ratio weight/capacity
		Model	2.5″	SCSI	(GB)	Year	(g rounded)		(g/GB)
2	Maxtor	7245SR			0.245	1993	15.2	4	62.04
3	Conner				0.635	1995	19.5	2	30.71
4	Fujitsu	MPF3204AT			20.4	2000	19	2	0.93
5	Seagate	ST320410A			20	2001	14.2	1	0.71
6	Quantum	ProDrive LPS			0.127	2002	21.3	4	167.72
7	Quantum	ST16AT			1.6	1996	13.6	2	8.50
8	Samsung	SV0432A			4.3	1999	20.5	2	4.77
10	Maxtor	32049H2			20.4	2000	13.4	1	0.66
12	Maxtor	32049H2			20.4	2000	13.4	1	0.66
13	WD	WD800			80	2002	11	2	0.14
14	WD Caviar	WD136AA			13.6	1999	19.7	2	1.45
15	WD	WD200			20	2000	2.5	2	0.13
16	WD	WD200			20	2001	2.5	2	0.13
17	WD Caviar	WDAC22100			21	1997	10.6	2	0.50
18	Seagate	ST8422A			8.6	1998	19.5	2	2.27
19	Seagate	ST3500418AS			500	2009	4.3	2	0.01
20	WD	307AA			30.7	2000	19.8	2	0.64
21	Maxtor	4D04H2			40	2002	11.3	1	0.28
22	Seagate	ST320423A			20.4	2000	15.9	1	0.78
23	Maxtor	4D04H2			40	2002	11.2	1	0.28
24	Maxtor	DiamondMax+8			40	2002	6.2	1	0.16
25	Seagate	ST340823A			40	2000	14.1	1	0.35
26	Maxtor	32049H2			20.4	2000	13.2	1	0.65

(continued)

V. Zepf, *Rare Earth Elements*, Springer Theses,
DOI: 10.1007/978-3-642-35458-8,

(continued)

Label	Manufacturer	Type	Capacity	Weight of magnets (sum)	Number of VCA magnets	Ratio weight/ capacity
27	Maxtor	4D04H2	20.4	2000 13.2	1	0.65
28	WD Caviar	32100	2	1996 10	1	5.00
29	Fujitsu	MPC3064AT	6.4	1998 21.4	2	3.34
30	WD	WD200	20	2000 2.6	2	0.13
31	Maxtor	32049H2	20.4	2000 13.2	2	0.65
32	WD	WD200	20	2000 2.5	2	0.13
33	Maxtor	32049H2	20.4	2000 13.4	2	0.66
34	WD	WD200	20	2000 2.6	2	0.13
36	Samsung	SP0411N	40	2005 3	2	0.08
37	Samsung	SP1654N	160	2006 15.7	2	0.10
38	Fujitsu	MPC3043AT	4.3	1998 21.4	2	4.98
39	WD	28400	8.5	1999 21	2	2.47
40	Fujitsu	MPB3021AT	2.1	1998 11.1	1	5.29
A1	Maxtor	80GB	80	2003 8.1	2	0.10
A5	Maxtor	4D040H2	40	2002 11.37	1	0.28
A6	WD	883D	500	2007 5.92	2	0.01
A7	Maxtor	N256	8.4	1999 21.09	2	2.51
A9	Maxtor	4D080H4	80	2002 11.35	1	0.14
A10	Maxtor	4D040H2	40	2005 11.42	1	0.29
A14	Quantum	Fireball K3	2.1	1998 9.24	1	4.40
A15	Quantum	Fireball Tm	2.1	1998 18.49	2	8.80
A16	Quantum	Fireball Ex	6.4	1998 13.13	2	2.05
A17	IBM SCSI	DMVS	9.1	1999 33.48	2	3.68
A18	WD	64AA	6.4	1999 19.49	2	3.05
A20	Maxtor	4D040H2	40	2002 11.26	1	0.28

(continued)

(continued)

Label	Manufacturer	Type			Capacity		Weight of magnets (sum)	Number of VCA magnets	Ratio weight/ capacity
A23	Maxtor	DiamondMax +8			40	2002	6.17	1	0.15
A27	Maxtor	4D040H2			40	2002	11.14	1	0.28
B1	Fujitsu	SCSI			300	2005	29.42	2	0.10
B2	Seagate	St51080A			1	2001	8.24	2	8.24
B3	Hitachi	Deskstar			164	2006	6.92	2	0.04
B4	Seagate	Barracuda			160	2006	8.12	2	0.05
B5	IBM 2.5″	DTCA24090	x		4	1998	4.1	2	1.03
B6	Seagate	Cheetah15K5 ST373455LC		x	73.4	2006	35.55	2	0.48
B7	Fujitsu	MAN3367MP SCSI			36.7	2001	21.48	2	0.59
B8	Fujitsu	MAN3735FC			73.4	2002	31.26	2	0.43
A3	WD	WD200			20	2000	2.56	2	0.13
A4	WD	WD100			10	2001	2.56	2	0.26
A8	WD	WD2300			20	2000	2.53	2	0.13
A24	WD	WD400			40	2003	2.56	2	0.06
A25	WD	WD200			20	2000	2.56	2	0.13
B11	Fujitsu 2,5″	MHN2150AT	x		15	2001	2.16	1	0.14
C1	Samsung	SP20A1H			20	2002	16.45	2	0.82
C2	Seagate	ST314220A			14.2	1999	19.57	2	1.38
C3	WD	WD2500			250	2005	11.12	2	0.04
C5	Seagate	ST330630A			30.6	1999	19.66	2	0.64
C6	Seagate	ST34321A			4.3	2003	14.13	2	3.29
C7	Seagate	ST33232A			3.2	1998	16.5	2	5.16
C8	IBM	DTTA 351010			10.1	1999	20.6	2	2.04
C9	WD	WDAC21200			1.3	1996	8.3	1	6.38

(continued)

(continued)

Label	Manufacturer	Type			Capacity		Weight of magnets (sum)	Number of VCA magnets	Ratio weight/ capacity
C10	Seagate	Momentus 5400.6	x		500	2009	2.6	2	0.01
C11	Maxtor	34098H4			40.9	2001	13.03	1	0.32
C12	Quantum FB	2110AT			2.1	1996	18.77	2	8.94
C13	IBM	DSAS 3540			0.5	1994	9.81	4	19.62
C14	WD	WD2500			250	2005	11.11	2	0.04
C15	Samsung	SV4002H			40	2001	16.44	2	0.41
C16	WD	2340			0.3	1994	12.1	4	40.33
C18	Seagate	ST39102LC		x	9.1	1994	29.4	2	3.23
C19	IBM	DDYS-T09170		x	9.1	2000	41.23	2	4.53
C20	IBM	DNES-309170			9	1999	29.12	2	3.24
C22	Quantum Dell	1620AT			1.6	1997	13.3	2	8.31
C25	Fujitsu	MAP3367NC SCSI		x	36.7	2003	21.44	2	0.58
C26	IBM	DPES31080			1	1995	9.29	2	9.29
C27	Fujitsu	MPB3043AT			4.3	1998	21.47	2	4.99
C28	Seagate	ST3300007LC			300	2005	26.31	2	0.09
C29	MIC Microscience Int'l Corp	HH1090 5.25″			0.08	1990	65.66		820.75
C30	Samsung	SV3002H			30	2001	16.38	2	0.55
C31	Seagate	ST3290A			0.261	1993	13.93	1	53.37
C32	Seagate	ST3660A			0.545	1995	14.73	1	27.03
C33	Seagate	ST3144A			0.131	1992	14.12	1	107.79
C34	Quantum	ProDrive ELS			0.17	1992	17.22	4	101.29
C35	Conner	CFP2105S			2.1	1995	27.44	2	13.07
C36	Maxtor	84320D4			4.3	1997	14.86	1	3.46

(continued)

(continued)

Label	Manufacturer	Type		Capacity		Weight of magnets (sum)	Number of VCA magnets	Ratio weight/ capacity
C37	Conner	CFP2105S		2.1	1998	27.18	2	12.94
C38	Quantum	1280AT		1.2	1995	10.16	2	8.47
C39	Seagate	ST52520A		2.5	1995	8.16	2	3.26
C40	Seagate	ST32550N		2.1	1996	17.39	2	8.28
C41	Seagate	ST38420A		8.6	1998	19.5	2	2.27
C42	Seagate	ST38410A		8.4	1999	15.43	1	1.84
C44	IBM	DNES309170	x	9	1999	29.44	2	3.27
C45	IBM	IC35L040AVER07		41	2001	27.66	2	0.67
C46	Quantum	GrandPrix		4.3	1995	71.25	2	16.57
C47	Seagate	ST39102LW	x	9.1	1997	29.27	2	3.22
C48	WD	64AA		6.4	1999	19.66	2	3.07
C49	Maxtor	34098H4		40.9	2001	13.01	1	0.32
C50	Seagate	ST3144A		0.131	1992	13.93	1	106.34
C51	Maxtor	D740Y-6L		40	2002	15.42	2	0.39
C52	Conner	CFA340A		0.345	1999	18.91	2	54.81
C53	Teac	SD340		0.043	1990	7.62	1	177.21
C54	IBM	DCHS	x	4.55	1996	34.82	4	7.65
C55	Maxtor	7345AT		0.345	1993	15.51	4	44.96
C56	Seagate	ST3390A		0.341	1993	17.54	2	51.44
C57	Maxtor	DiamondMax D		200	2004	14.36	2	0.07
C58	WD	WD1200		120	2002	11.2	2	0.09
C59	WD	WD1200		120	2003	11.2	2	0.09
C60	Maxtor	7540AQ		0.54	1995	7.63	2	14.13
C61	Seagate	ST1181677LCV	x	181	2001	76.55	2	0.42
C62	Samsung	SP0802N		80	2004	15.82	2	0.20

(continued)

(continued)

Label	Manufacturer	Type			Capacity		Weight of magnets (sum)	Number of VCA magnets	Ratio weight/capacity
C63	Seagate	ST34321A			4.3	1997	14	2	3.26
C64	Maxtor	DM9 YAR41BWO			80	2004	8	2	0.10
C65	Fujitsu	MAR2020AT	x		20	2002	2.48	1	0.12
C66	IBM	IC25N020ATCS04-0	x		20	2002	2.93	1	0.15
C67	Hitachi	DK23EB-40	x		40	2002	2.78	1	0.07
C68	Fujitsu	MAR2030AT	x		30	2001	2.43	1	0.08
C69	Toshiba	MK8113MAT	x		8.2	1999	2.57	1	0.31
J1	Seagate	ST313030A			13	1998	19.32	2	1.49
J2	Seagate	ST340016A			40	2002	8.79	2	0.22
J3	Compaq	BD03674555		x	36.4	2002	18.41	2	0.51
J4	Seagate	ST34321A			4.3	1997	13.96	2	3.25
J5	Maxtor	7540AQ			0.54	1995	7.39	2	13.69
J6	Seagate	ST39173W		x	9	1997	27.26	2	3.03
J7	Seagate	ST314220A			14.2	1999	19.55	2	1.38
J8	Seagate	ST31276A			1.28	1996	19.6	1	15.31
J9	IBM	DCAS 32160			2	1997	21.37	2	10.69
J10	Seagate	ST320011A			20	2001	8.75	2	0.44
J12	Compaq	MAA3182SC		x	18.2	1998	39.05	2	2.15
41	Quantum	BigFootx			2.5	1999	22.14	2	8.86

Notes

Entries crossed out were not taken into analysis due to inadequate data availability or validity

HDDs not labeled 2.5″ were of 3.5″ size

5.25″ HDDs were also excluded as well

2.5″ HDDs were considered without annual distinction due to low sample quantities

Annex D
REE Uses and Applications

REE uses and application

Element	Supply[a] (% REO)	Demand[a] (kt REO)	Uses	(s)
Mixed REO			**Misch metal (as natural occurrence in bastnaesite and monazite)**	a
			Steels (refining)	
			Ductile iron	
			Lighter flints	
			Alloying agent (Mg, Al, cast iron)	
			NiMH batteries	
			Petroleum cracking catalyst	a
			Polishing compounds	a
			Fertilizers	a
Lanthanum	25	36	**Nickel metal hydride batteries (Prius, forklifts)**	a
La			Hydrogen storage alloys $LaNi_3$	a
			Alloying Agent—Mg, Al, cast iron, superalloys	a
			Sputtering targets	a
			Optical	a
			Lenses (increase index of refraction)	
			Host for phosphors	a
			Fluorescent lamps	
			X-ray intensifying screens	
			Green phosphors	b
			Petroleum cracking catalyst	a
			Condensers, $BaTiO_3$	a
			Heavy metal fluoride glasses	a
			Fiber optics	
			Night Vision Goggles	a
			Cathode material in SOFC (Solid Oxide fuel cell)	b
			Laser crystals based on yttrium-lanthanum-fluoride (YLF) type	b
Cerium	30	42 demand	**Catalysts**	a
Ce		55 supply	**Automotive 3-way-emission catalysts**	
			Petroleum cracking	
			Water purification	
			Glass additives	a, b
			Decolorizer, Opacifier (keeps iron its ferrous state)	
			UV light absorption (protection of contents from deterioration; medical glassware, aviation glass, food glassware)	

(continued)

V. Zepf, *Rare Earth Elements*, Springer Theses,
DOI: 10.1007/978-3-642-35458-8,

(continued)

REE uses and application

Element	Supply[a] (% REO)	Demand[a] (kt REO)	Uses	(s)
			Polishing media	a
			Glass, lenses, face plates of monitors	
			Semiconductors	
			Ce-stabilized ZrO_2	a
			Structural ceramics	
			Synthetic gems (pale yellow)	
			Phosphors	a
			Fluorescent lamps (e.g. Auerlight)	
			Scintillation counters	
			Condensers, $BaTiO_3$	a
			Alloying agent—Mg, Al, cast iron, superalloys	a
			Steels (refining)	a
			Ceramics	b
			Dental compositions	
Praseo-dymium Pr	5	7	**Additive to $Nd_2Fe_{14}B$ (reduces amount of Nd used)**	a
			Pr-stabilized ZrO_2 (synthetic gems—light green)	a
			Coloring agents	a
			Glass—light green	
			Ceramic tile—yellow	
			Condensers, $BaTiO_3$	a
			Scintillators forx-ray tomography	a
			Glass blower's and welder's goggles (with Nd)	a
			Telecommunication systems	b
			Dopant in fluoride fibers	
Neodymium, Nd	16	23 tight, slight shortage, 2.8kt Nd for WTG (USA)	**$Nd_2Fe_{14}B$ permanent magnets**	a
			Electric motors (largest), spindles for computer hard drives (second largest), cell phones, iPods ®, direct drive, wind turbines, actuators	
			Alloying agent for Mg alloys	a
			Lasers	a
			Glass, YAG	b
			Medical, drilling, welding, material processing	
			Metal halide lamps	a
			Nd-stabilized ZrO_2 synthetic gems	a
			Pink, purple	

(continued)

(continued)

REE uses and application

Element	Supply[a] (% REO)	Demand[a] (kt REO)	Uses	(s)
			Coloring agent	a
			Glass (pink to purple)	
			Condensers, $BaTiO_3$ (?)	a
			Glass blower's and welder's goggles (with Pr)	a
Samarium	2	2.8	**SmCo permanent magnets for above room temperature applications**	a
Sm			Coloring agent	a
			Glass (light yellow)	
			Phosphors	a
			Monitors, TV screens	
			Nuclear industry—radiation shielding	b
			Lasers (dielectric properties)	b
			Coatings and capacitors at microwave frequencies	
Europium	0.3	2.1	**Phosphors (red colors)**	a
Eu			**TVs**	
			Fluorescent lamps (CFL and LFL)	
			Scintillators for x-ray tomography	
			Phosphors (blue color)	a
			LEDs	a
			Nuclear industry—radiation shielding	b
Gadolinium	1.5	2.1	**Host for phosphors**	a
Gd			**Fluorescent lamps**	
			Scintillators for x-ray tomography	
			MRI contrast agents	a
			Nuclear— fuel rod addition, safety	a
			X-ray intensifying screen	a
			YGG (Yttrium Gadolinium Garnet)	a
			Communications (radio frequencies)	
			Phase shifters, tuners, filters, radar	
			Optical lenses	a
			Magnetooptical materials	b
Terbium	0.3	0.4 tight, short	**Phosphors (green)**	a
Tb			**Fluorescent lamps (CFL and LFL)**	
			Monitors and TV screens	
			LEDs	
			X-ray intensifying screens	a
			Terfenol-D (Tb_xDy_y)Fe_2	a
			Magnetrorestrictive alloy	
			Magneto-optic disks	a

(continued)

(continued)

REE uses and application

Element	Supply[a] (% REO)	Demand[a] (kt REO)	Uses	(s)
Dysprosium Dy	1.0	1.4 Tight short	**Additive to $Nd_2Fe_{14}B$ permanent magnets to improve high temperature performance, increase coercivity**	a
			Phosphors	a
			Nuclearindustry—radiationshielding	b
			Ceramics composition based on BaTiO types	b
Holmium	0.1	0.1 surplus	Research	a
Ho			Metal halide lamps	a
			YIG (Yttrium- iron-garnet) lasers for microwave equipment	b
			YAG and YLF solid state lasers	
Erbium	0.5	0.7	**Fiber optics**	a
Er			**Signal amplifiers**	
			Lasers (mainly medical /surgical and dental use)	a, b
			Coloring agent (pink)	a
			Eyewear	
			Decorative glassware	
			Er-stabilized ZrO2 synthetic gems (pink)	a
Thulium	0.1	0.1 Surplus	X-ray intensifying screens	a
Tm			Metal halide lamps	a
			Crystal and laser making	, b
			Research	
Ytterbium	0.2	0.3		a
Yb		Surplus	Optical lenses	
			Pressure sensors (metal)	a
			Research	a
			Fiber amplifier and fiber optic technologies	b
			Ceramics	b
			Exhaust oxygen sensors (substitute for Y)	
Lutetium	0.1	0.1	Research	a
Lu		Surplus	Host for scintillator detectors and X-ray phosphors	a
			Optical lenses	a

(continued)

(continued)

REE uses and application

Element	Supply[a] (% REO)	Demand[a] (kt REO)	Uses	(s)
Yttrium	6	8.5	**Host for phosphors**	a
Y			**Fluorescent lamps (CFL and LFL)**	
			Monitors and TV and LCD screens	
			YAG laser host material	a
			Y-stabilized ZrO_2	a
			Electronic (oxygen sensor, automobiles)	
			Structural (rough (?) ceramics)	
			Synthetic gems (simulated diamonds)	
			Thermal barrier coatings (turbines, aircraft engines)	
			YGG	a
			Communication …	
			YIG Yttrium Iron Garnet	a
			Communication (radio frequencies)	
			Radars, phase shifters, filters	
			YBa_2CuO_2 high temperature superconductor	a
			Power transmission lines	
			Magnets for wind turbines	
			Ceramics—sintering (Y_2O_2) (?)	a
			Optical lenses	a
			Alloying agent for	a
			Superalloys …	
			Mg high temp creep resistance	
Scandium			Ceramics	b
Sc			Lasers	
			Phosphors	
			High performance alloys	

Notes S—data sources, *a*—Gschneidner 2011, *b*—Gupta and Krishnamurthy 2005
Bold letters indicate major applications
Supply and demand data are just copied from A. More details about data reliability are given in the text
Not available but very important are the respective quantitative shares and the risk for dissipation

About the Author

Volker Zepf, born in 1967 in Rottweil, Germany, is a scientific employee at Augsburg University, Chair of Resource Strategy. He studied Geography at the University of Augsburg with emphasize on Human Geography and Geoinformatics, and as secondary topics Geology, Resource Geography, Regional Planning and Ecological Business Management. His major research topics are raw materials in general and rare metals in particular. The geographical aspects space and time play a vital role in the analysis of the raw material issues. This is also represented in his Master Thesis 'Africa in neocolonialistic times– about the importance of the African raw materials for key technologies of the globalized world' (Thesis written in German language). Next to the task as a researcher he is Coordinator of the Graduate Program of the Institute for Materials and Resource Management at Augsburg University.

V. Zepf, *Rare Earth Elements*, Springer Theses,
DOI: 10.1007/978-3-642-35458-8,

MIX
Papier aus verantwortungsvollen Quellen
Paper from responsible sources
FSC® C105338

If you have any concerns about our products,
you can contact us on
ProductSafety@springernature.com

In case Publisher is established outside the EU,
the EU authorized representative is:
Springer Nature Customer Service Center GmbH
Europaplatz 3, 69115 Heidelberg, Germany

Printed by Libri Plureos GmbH
in Hamburg, Germany